Working from Within

Working from Within

The Nature and Development of Quine's Naturalism

SANDER VERHAEGH

OXFORD
UNIVERSITY PRESS

OXFORD
UNIVERSITY PRESS

Oxford University Press is a department of the University of Oxford. It furthers the University's objective of excellence in research, scholarship, and education by publishing worldwide. Oxford is a registered trade mark of Oxford University Press in the UK and certain other countries.

Published in the United States of America by Oxford University Press
198 Madison Avenue, New York, NY 10016, United States of America.

Library of Congress Cataloging-in-Publication Data
Names: Verhaegh, Sander, author.
Title: Working from within : the nature and development of Quine's naturalism /
Sander Verhaegh.
Description: New York, NY, United States of America : Oxford University Press, [2018] |
Includes bibliographical references and index.
Identifiers: LCCN 2018016072 (print) | LCCN 2018028844 (ebook) |
ISBN 9780190913168 (online content) | ISBN 9780190913175 (updf) |
ISBN 9780190913182 (epub) | ISBN 9780190913151 (cloth : alk. paper)
Subjects: LCSH: Quine, W. V. (Willard Van Orman) | Naturalism.
Classification: LCC B945.Q54 (ebook) | LCC B945.Q54 V47 2018 (print) |
DDC 191—dc23
LC record available at https://lccn.loc.gov/2018016072

1 3 5 7 9 8 6 4 2

Printed by Sheridan Books, Inc., United States of America

It is understandable [. . .] that the philosopher should seek a transcendental vantage point, outside the world that imprisons [the] natural scientist and mathematician. He would make himself independent of the conceptual scheme which it is his task to study and revise. "Give me πoυ στω [a place to stand]," Archimedes said, "and I will move the world." However, there is no such cosmic exile. [. . .] The philosopher is in the position rather, as Neurath says, "of a mariner who must rebuild his ship on the open sea."

—W. V. Quine (notes for *Sign and Object*, November 5, 1944)

CONTENTS

PREFACE

This book is the result of almost nine years of thinking and writing about Quine's naturalism. The project started with a term paper on the Carnap-Quine debate for Jeanne Peijnenburg's inspiring course on the history of analytic philosophy in 2008 and gradually evolved into a dissertation about the relation between holism and naturalism in Quine's philosophy, which I defended in Groningen in 2015. Encouraged by the warm support of my colleagues and a number of fellow Quine scholars, I spent the last three years revising and extending the manuscript. Having acquired a serious case of "archive fever," a significant chunk of this period was spent at the W. V. Quine Papers at Houghton Library, Harvard University. Using his published corpus as well as thousands of letters, notes, draft manuscripts, lectures, grant proposals, teaching materials, and annotations collected at the archives, I have aimed to write a book that reconstructs both the nature and the development of Quine's naturalism. The notes, letters, and lectures that have most influenced my views about (the development of) Quine's naturalism are transcribed and collected in the appendix.

Parts of this book have been published elsewhere. I thank the respective journals for granting me permission to reprint these papers:

Quine's argument from despair (2014). *British Journal for the History of Philosophy, 22*(1), 150–73. (Chapter 2)

Blurring boundaries: Carnap, Quine, and the internal-external distinction (2017). *Erkenntnis, 82*(4), 873–90. (Chapter 3)

Boarding Neurath's boat: The early development of Quine's naturalism (2017). *Journal for the History of Philosophy, 55*(2), 317–42. (Sections 4.1–4.2, 6.2, and 6.6)

Quine on the nature of naturalism (2017). *Southern Journal of Philosophy, 55*(1), 96–115. (Section 4.5)

Sign and object: Quine's forgotten book project (forthcoming). *Synthese*. https://doi.org/10.1007/s11229-018-1693-z. (Chapter 5)

Quine's "needlessly strong" holism (2017). *Studies in History and Philosophy of Science, Part A, 61*(1), 11–20. (Sections 6.8–6.11)

Setting sail: The development and reception of Quine's naturalism (forthcoming). *Philosophers' Imprint*. (Chapter 7)

Still, this book aims to be more than a collection of papers. Many sections have been revised, extended, and rewritten for the purposes of this book. Moreover, an introductory chapter, a conclusion, several new sections (i.e., 4.3–4.4, 4.6–4.8, 5.5, 6.3–6.5, 6.7), and an appendix have been included to fill the gaps and to present a comprehensive account of Quine's naturalism.

Much of this research has been funded by the Netherlands Organisation for Scientific Research (grants 322-20-001 and 275-20-064). My research visits to Harvard University, Houghton Library, Bethel, Connecticut, and the Harvard University Archives have been funded by a Kristeller-Popkin Travel Fellowship from the *Journal of the History of Philosophy*, a Rodney G. Dennis Fellowship in the Study of Manuscripts from Houghton Library, and a Travel Bursary from the Evert Willem Beth Foundation. I am very grateful for this financial support.

Finally, it is my great pleasure to thank the many people without whom I could not have completed this book: Jeanne Peijnenburg, Allard Tamminga, Lieven Decock, Gary Ebbs, Hans-Johann Glock, Peter Hylton, Gary Kemp, Fred Muller, Martin Lenz, and two anonymous referees for Oxford University Press who commented on earlier versions of this book; Harvard's Department of Philosophy, Nyasha Borde, Monique Duhaime, Juliet Floyd, Warren Goldfarb, Leslie Morris, Mark Richard, Thomas Ricketts, and the staff at Houghton Library and the Harvard University Archives for their warm welcome during my research visits; Douglas and Maryclaire Quine for welcoming me into their house to examine the remaining fifty-seven boxes of unprocessed archive material; Hannah Doyle, Lucy Randall, Richard Isomaki, Leslie Johson, and Tharani Ramachandran at Oxford University Press and Newgen for smoothly guiding my manuscript through the editing and production process; Jody Azzouni, Richard Creath, Fons Dewulf, Dagfinn Føllesdal, James Levine, Eric Schliesser, Andrew Smith, Serdal Tumkaya, Thomas Uebel, Wim Vanrie, the members of the EPS seminar in Tilburg, the members of the PCCP and the WiP seminars in Groningen, and audiences in Amsterdam, Athens, Bloomington, Bologna, Calgary, Chicago, Denver, Edinburgh, Groningen, Konstanz, Manchester, Milan, Modena, Munich, Rotterdam, Tampa, Tilburg, and Zurich for their comments on various papers and chapters in various stages; Reto Gubelmann, Frederique Janssen-Lauret, Gary Kemp, and Sean Morris for their wonderful Quine workshops in Denver, Glasgow, Manchester, and Zurich; and Douglas B. Quine (W. V. Quine Literary Estate) and Catherine Z. Elgin (Literary executor for Nelson Goodman) for granting me permission to publish some of Quine's and Goodman's papers, notes, and letters in an appendix to this book. Of course, the above individuals do not necessarily agree with my conclusions, and they are not responsible for any residual errors or omissions.

A note on citation and transcription: Unless specified differently, the unpublished documents I refer to in this book are part of the W. V. Quine Papers, collection MS Am 2587, Houghton Library, Harvard University. In the main text and in the footnotes, I refer to these documents by citing dates (if known) and item numbers. A letter of Quine to the American Philosophical Association, for example, could be referred to as (January 1, 1950, item 31). For quick reference, the item numbers are listed in the list of abbreviations. The items' full titles and box numbers are provided in the bibliography. In transcribing Quine's autograph notes, drafts, and letters, I have aimed to minimize editorial interference and chosen not to correct ungrammatical shorthand. In referring to Quine's *published* work, I use abbreviations (listed in the list of abbreviations) as well as the year in which the paper or book was first published. If the paper is incorporated in one of Quine's collections of papers, the page numbers will refer to this collection. The first page of "Two Dogmas of Empiricism," for example, will be referred to as (TDE, 1951a, 20). The details of Quine's published work cited in this book are provided in the bibliography.

In his autobiography, *The Time of My Life*, Quine argues that science and history of science appeal to very different tempers: "An advance in science resolves an obscurity, a tangle, a complexity, an inelegance, that the scientist then gratefully dismisses and forgets. The historian of science tries to recapture the very tangles, confusions, and obscurities from which the scientist is so eager to free himself" (TML, 1985a, 194). I hope this book accomplishes a bit of both; although I will show that Quine faced many tangles, confusions, and obscurities in developing his naturalism, I also hope to convey that the system he created is one of great clarity and elegance.

Sander Verhaegh
June 2018

ABBREVIATIONS

This section provides a list of abbreviations used in this book to refer to W. V. Quine's published and unpublished work. Detailed references can be found in the bibliography. A complete list of Quine's publications up to approximately 1997 (including reprints and translations) can be found in Yeghiayan (2009). For an overview of papers, books, reprints, and translations after 1997, see Douglas B. Quine's website http://www.wvquine.org. Most of Quine's unpublished papers, letters, lectures, and notebooks can be accessed at Houghton Library, Harvard University. The documents' call numbers, box numbers, and item numbers are provided in the bibliography. A finding aid for much of Quine's unpublished work can be found at http://oasis.lib.harvard.edu/oasis/deliver/~hou01800. Archival sources from Quine's unprocessed papers, Quine's library, and the papers of Nelson Goodman are not listed here because they are not properly itemized. Details about these sources are listed in the bibliography.

Archival Sources

Item 31	Correspondence with the American Philosophical Association (1936–1986)
Item 40	Correspondence Ap- through As- (various dates)
Item 86	Correspondence with Lars Bergström (1988–1996)
Item 96	Correspondence with Evert Willem Beth (1947–1964)
Item 205	Correspondence with the Center for Advanced Study in the Behavioral Sciences (1955–1979)
Item 224	Correspondence with Alonzo Church (1935–1994)
Item 231	Correspondence with Joseph T. Clark (1951–1953)
Item 234	Correspondence Co- (various dates)
Item 248	Correspondence with Columbia University (1949–1970)

Item 254	Correspondence with James Bryant Conant (1951–1979)
Item 260	Correspondence with John Cooley (1932–1962)
Item 270	Correspondence with Richard Creath (1977–1998)
Item 287	Correspondence with Donald Davidson (1957–1997)
Item 293	Correspondence with Grace De Laguna (1950–1954)
Item 306	Correspondence Di- through Do- (various dates)
Item 315	Correspondence with Burton Dreben (1948–1997)
Item 336	Correspondence Er- through Ez- (various dates)
Item 370	Correspondence with Philipp Frank (1951–1967)
Item 420	Correspondence with Nelson Goodman (1935–1994)
Item 471	Correspondence with Harvard University, Department of Philosophy (1930–1994)
Item 473	Correspondence with Harvard University, Faculty of Arts and Sciences (1931–1998)
Item 475	Correspondence with Harvard University, Grants (1941–1988)
Item 479	Correspondence with Harvard University, President's Office (1937–1998)
Item 499	Correspondence with Carl Gustav Hempel (1936–1997)
Item 529	Correspondence with Christopher Hookway (1988)
Item 530	Correspondence with Paul Horwich (1991–1992)
Item 545	Correspondence with the Institute for Advanced Study (1955–1965)
Item 553	Correspondence Ir- through Iz- (various dates)
Item 570	Correspondence with the *Journal of Symbolic Logic* (1936–1996)
Item 616	Correspondence with Imre Lakatos (1964–1974)
Item 637	Correspondence with Michele Leonelli (1966–1998)
Item 643	Correspondence with Clarence Irving Lewis (1929–1996)
Item 675	Correspondence with Joseph Margolis (1967 and undated)
Item 724	Correspondence with Hugh Miller (1948–1952)
Item 741	Correspondence with Charles W. Morris (1936–1947)
Item 755	Correspondence with John Myhill (1943–1985)
Item 758	Correspondence with Ernest Nagel (1938–1964)
Item 885	Correspondence with Hilary Putnam (1949–1993)
Item 921	Correspondence with the Rockefeller Foundation (1945–1980)
Item 958	Correspondence Sc- (various dates)
Item 972	Correspondence with Wilfrid Sellars (1938–1980)
Item 1001	Correspondence with B. F. Skinner (1934–1998)
Item 1005	Correspondence with J. J. C. Smart (1949–1998)

Item 1014	Correspondence with Ernest Sosa (1970–1995)
Item 1200	Correspondence with Paul Weiss (1937–1972)
Item 1213	Correspondence with Morton White (1939–1998)
Item 1221	Correspondence with Donald Cary Williams (1940–1994)
Item 1237	Correspondence with Joseph Henry Woodger (1938–1982)
Item 1239	Correspondence with the World Congress of Philosophy (1952–1998)
Item 1244	Correspondence with Morton G. Wurtele (1938–1997)
Item 1263	Correspondence concerning requests to publish or for copies (1950–1959)
Item 1355	Editorial correspondence with D. Reidel Publishing Company (1960–1982 and undated)
Item 1391	Editorial correspondence concerning *Mathematical Logic* (1939–1940)
Item 1401	Editorial correspondence concerning *Methods of Logic* (1947–1950)
Item 1422	Editorial correspondence concerning *Ontological Relativity* (1968–1988)
Item 1423	Editorial correspondence concerning *On Translation* (1956–1959)
Item 1443	Editorial correspondence concerning On What There Is (1948–1975)
Item 1467	Editorial correspondence concerning *Theory of Deduction* (1946–1949)
Item 1488	Editorial correspondence concerning *Word and Object* (1952–1960)
Item 1489	Editorial correspondence concerning *Word and Object* (1959–1980 and undated)
Item 1490	Editorial correspondence concerning *Words and Objections* (1966–1974)
Item 2388a	Quine's annotated copy of Putnam's *Meaning and the Moral Sciences* (undated)
Item 2441	Epistemology Naturalized; or, the Case for Psychologism (1968)
Item 2498	In conversation: Professor W. V. Quine. Interviews by Rudolf Fara (1993)
Item 2733	Russell's Ontological Development (1966–1967)
Item 2756	Stimulus and Meaning (1965)
Item 2829	*A Short Course on Logic* (1946)
Item 2830	*Theory of Deduction* (1948)

Item 2836	Foreword to the Third Edition of *From a Logical Point of View* (1979)
Item 2884	Questions for Quine by Stephen Neale (1986)
Item 2902	For Rockefeller Lecture (1968)
Item 2903	For symposium with Sellars (1968)
Item 2928	The Inception of "New Foundations" (1987)
Item 2948	Kinds (1967)
Item 2952	Levine seminar questions for Quine (1982)
Item 2954	Logic, Math, Science (1940)
Item 2958	Mathematical Entities (1950)
Item 2969	Nominalism (1937)
Item 2971	The Notre Dame Lectures (1970)
Item 2994	Ontological Relativity (1967)
Item 2995	Ontological Relativity (1968)
Item 3011	The Place of a Theory of Evidence (1952)
Item 3015	The Present State of Empiricism (1951)
Item 3102	Word and Object Seminar (1967)
Item 3158	Philosophy 148 (1953)
Item 3169	Early Jottings on Philosophy of Language (1937–1944)
Item 3170	Erledigte Notizen (various dates)
Item 3181	Ontology, Metaphysics, etc. (1944–1951)
Item 3182	Philosophical notes (various dates)
Item 3184	Pragmatism, etc. (1951–1953)
Item 3225	Miscellaneous papers (1925–1931)
Item 3236	Papers in philosophy (1930–1931)
Item 3254	General Report of My Work as a Sheldon Traveling Fellow 1932–1933 (1934)
Item 3266	Philosophy 148 (ca. 1947)
Item 3277	Oxford University Lectures (1953–1954)
Item 3283	Oxford University Lecture: Philosophy of Logic (1953–1954)

Published Works

ANM	Animadversion on the Notion of Meaning (1949)
AWVQ	Autobiography of W. V. Quine (1986)
CA	Carnap (1987)
CB	Comment on Berger (1990)
CCE	*Confessions of a Confirmed Extensionalist and Other Essays* (2008)

CD	Contextual Definition (1995)
CGC	A Comment on Grünbaum's Claim (1962)
CLT	Carnap and Logical Truth (1954)
CNT	Comments on Neil Tennant's "Carnap and Quine" (1994)
CPT	Carnap's Positivistic Travail (1984)
CVO	Carnap's Views on Ontology (1951)
DE	Designation and Existence (1939)
EBDQ	Exchange between Donald Davidson and W. V. Quine Following Davidson's Lecture (1994)
EESW	On Empirically Equivalent Systems of the World (1975)
EN	Epistemology Naturalized (1969)
EQ	Existence and Quantification (1968)
FHQP	Four Hot Questions in Philosophy (1985)
FLPV	*From a Logical Point of View* (1953/1961)
FM	Facts of the Matter (1977)
FME	Five Milestones of Empiricism (1975)
FSS	*From Stimulus to Science* (1995)
GQW	Nelson Goodman, W. V. Quine and Morton White: A Triangular Correspondence (1947)
GT	Grades of Theoreticity (1970)
IA	Autobiography of W. V. Quine (1986)
IOH	Identity, Ostension, Hypostasis (1946)
IQ	The Ideas of Quine. Interview by B. Magee (1978)
ITA	Indeterminacy of Translation Again (1987)
IV	Immanence and Validity (1991)
IWVQ	Interview with Willard Van Orman Quine. Interview by L. Bergström and D. Føllesdal (1994)
LAOP	A Logistical Approach to the Ontological Problem (1939)
LC	Lectures on Carnap (1934)
LDHP	Lectures on David Hume's Philosophy (1946)
LP	Linguistics and Philosophy (1968)
ML1	*Methods of Logic*. First edition (1950)
ML4	*Methods of Logic*. Fourth edition (1982)
MSLT	Mr. Strawson on Logical Theory (1953)
NEN	Notes on Existence and Necessity (1943)
NK	Natural Kinds (1969)
NLOM	Naturalism; Or, Living within One's Means (1995)
NNK	The Nature of Natural Knowledge (1975)
NO	Nominalism (1946)
OI	Ontology and Ideology (1951)
OME	On Mental Entities (1953)

ONAS	On the Notion of an Analytic Statement (1946)
OR	Ontological Relativity (1968)
ORE	*Ontological Relativity and Other Essays* (1969)
ORPC	Ontological Remarks on the Propositional Calculus (1934)
ORWN	Ontological Reduction and the World of Numbers (1964)
OW	Otherworldly (1978)
OWTI	On What There Is (1948)
PL	*Philosophy of Logic* (1970/1986)
PPE	The Pragmatists' Place in Empiricism (1975)
PPLT	Philosophical Progress in Language Theory (1970)
PR	Posits and Reality (1955)
PT	*Pursuit of Truth* (1990/1992)
PTF	Progress on Two Fronts (1996)
QCC	The Quine-Carnap Correspondence (1932–1970)
QD	*Quine in Dialogue* (2008)
QSM	Quine Speaks His Mind. An interview by E. Pivcevic (1988)
QU	*Quiddities* (1987)
QWVO	Quine/'kwain/, Willard Van Orman (b. 1908) (1996)
RA	Responses to Articles by Abel Bergström, Davidson, Dreben, Gibson, Hookway, and Prawitz (1994)
RC	Reply to Chomsky (1968)
RCP	Reply to Charles Parsons (1986)
RE	Reactions (1995)
RES	Responses to Szubka, Lehrer, Bergström, Gibson, Miscevic, and Orenstein (1999)
RGE	Response to Gary Ebbs (1995)
RGH	Reply to Geoffrey Hellman (1986)
RGM	Responding to Grover Maxwell (1968)
RHP	Reply to Hilary Putnam (1986)
RJV	Reply to Jules Vuillemin (1986)
RMW	Reply to Morton White (1986)
ROD	Russell's Ontological Development (1966)
RR	*The Roots of Reference* (1973)
RRN	Reply to Robert Nozick (1986)
RS1	Reply to Stroud (1968)
RS2	Reply to Stroud (1981)
RSM	Reply to Smart (1968)
RST	Reply to Stenius (1968)
RTD	Reply to Davidson (1968)
RTE	Responses to Essays by Smart, Orenstein, Lewis and Holdcroft, and Haack (1997)

SCN	Steps toward a Constructive Nominalism (1947)
SLP	*Selected Logic Papers* (1966/1995)
SLS	The Scope and Language of Science (1954)
SM	Stimulus and Meaning (1965)
SN	Structure and Nature (1992)
SSS	The Sensory Support of Science (1986)
STL	*Set Theory and its Logic* (1963/1969)
TC	Truth by Convention (1936)
TDE	Two Dogmas of Empiricism (1951)
TDR	Two Dogmas in Retrospect (1991)
TI	Three Indeterminacies (1990)
TML	*The Time of My Life* (1985)
TR	Truth (1994)
TT	*Theories and Things* (1981)
TTPT	Things and Their Place in Theories (1981)
VD	Vagaries of Definition (1972)
VITD	On the Very Idea of a Third Dogma (1981)
WB	*The Web of Belief* (with J. S. Ullian 1970/1978)
WDWD	Where Do We Disagree (1999)
WIB	What I Believe (1984)
WO	*Word and Object* (1960)
WP	*The Ways of Paradox and other essays* (1966/1976)
WPB	What Price Bivalence? (1981)
WRML	Whitehead and the Rise of Modern Logic (1941)
WW	The Way the World Is (1986)

1

Introduction

1.1. Historical Background

How are we to characterize the relation between science and philosophy? Is there a distinctive philosophical method which leads us to knowledge that cannot be obtained in the sciences? Is philosophy the "queen of the sciences"? Or is philosophy itself a branch of science such that its inquiries are to be integrated with investigations in, for example, mathematics, physics, and psychology?

Questions like these have been at the forefront of philosophical debate in the past few centuries. In answering them, philosophers have often pictured their subject as a foundational discipline—aiming to provide, inspect, and develop the foundations of science. Especially in metaphysics and epistemology, the foundationalist picture has proven to be influential. Traditional epistemologists have set out to answer the question whether our best scientific theories are *truly* justified, and traditional metaphysicians have sought to answer the question whether the entities posited by those theories *really* exist. If such foundational questions are to be answered in a satisfying way, these philosophers argued, we cannot presuppose any scientific knowledge in our philosophical inquiries. After all, it would be viciously circular to rely on science in vindicating science.

Naturalists reject these ideas about the relation between science and philosophy. In metaphysics, naturalists argue that there is no distinctively philosophical notion of "reality" that is somehow more fundamental than the concept we use in our everyday inquiries. Epistemologically, naturalists argue that scientific knowledge cannot be validated from an external (extrascientific) perspective; our best scientific theories are justified in an everyday empirical sense, and there is no additional, distinctively philosophical question about their justification. Instead, naturalists argue that philosophy and science are *continuous*; although philosophers have particular interests and methods—just as, for example, chemists and sociologists have distinctive aims and approaches—they are participants in the scientific enterprise at large.

Naturalism has a rich and complicated history. Although it is probably the first time that so many philosophers identify themselves as naturalists, it would be a mistake to think of the position itself as relatively new; naturalistic pictures of inquiry, at least in some varieties, are almost as old as science itself. Initially, naturalists were primarily involved in debates about the *supernatural*—i.e. in debates about the question whether reality is exhausted by nature as it is studied by the sciences. In response to the predominantly theistic cosmologies of their time, naturalists defended the view that we do not require supernatural laws, forces, and entities in explaining the workings of the world.[1]

Naturalistic worldviews gained prominence especially in nineteenth-century Europe. In the wake of Darwin's work on evolution, the early development of psychology as an independent discipline, and the steady advances in mathematics, physics, and chemistry, the prospects of a complete, naturalistic perspective on reality became a frequently debated subject among philosophers and scientists. In these often heated discussions, naturalists called for scientific explanations of supposedly supernatural phenomena, i.e. for a picture of reality in which man, mind, and morality are conceived as part of the natural order.[2]

Across the Atlantic, naturalism came to full bloom from the early 1920s onwards when a group of pragmatist philosophers, among whom John Dewey, Roy Wood Sellars, George Santayana, and, somewhat later, Ernest Nagel, Charles Morris, and Sidney Hook, came to develop views they identified as naturalistic.[3] Most of them defended a two-sided naturalism; they argued for the *metaphysical* claim that reality is exhausted by nature as it is studied by the sciences, thus dismissing appeals to supernatural explanation; and they defended the *methodological* claim that scientific method is the most reliable route to knowledge, thereby suggesting that philosophers ought to adopt this method.[4]

[1] Indeed, the *Oxford English Dictionary* (*OED*) still defines "philosophical naturalism" as a "view of the world, and of man's relation to it, in which only the operation of natural (as opposed to supernatural or spiritual) laws and forces is admitted." The *OED* dates the term back to 1750, when William Warburton described Amiannus Marcellinus as "a religious Theist [. . .] untainted with the Naturalism of Tacitus." Before the eighteenth century, the term "naturalist" was primarily used to refer to "experts in" or "students of" what we would now call natural science.

[2] See, for example, Spencer (1862), Lewes (1874), Huxley (1892), and Haeckel (1899). Broadly naturalistic philosophies were also developed by materialists, empiricists, and, to a certain extent, positivists. See Büchner (1855), Mill (1865), and Comte (1830–1842). For some influential antinaturalist responses from theistic, idealistic, and analytic philosophers, see Balfour (1895), Ward (1896), and Moore (1903).

[3] See Sellars (1922), Santayana (1923), Dewey (1925), and Nagel (1954), as well as the essays collected in Krikorian (1944) and Ryder (1994). For a history of naturalism in the United States before the 1920s, see Larrabee (1944).

[4] See Sellars (1924), Randall (1944), and, for a historical discussion, Kim (2003). Today, it is still quite common to distinguish between metaphysical and methodological varieties of naturalism. See Moser and Yandell (2000, §1), De Caro and Macarthur (2004b, pp. 2–8), and Papineau (2009).

Both branches of naturalism are compatible with the claim that science and philosophy are continuous, however. For if one accepts that philosophers are participants in the scientific enterprise at large, both one's metaphysical views and one's ideas about philosophical methodology might require revision.

The most influential naturalist of this group was John Dewey, a pragmatist who denied that there is any distinctively philosophical perspective on reality. Dewey, himself building on the antifoundationalism of predecessors like Peirce and James, viewed mind as part of the world that is studied by the sciences. For him, the philosopher is someone who "has no private store of knowledge or methods for attaining truth" and must therefore utilize "the best available knowledge of its time and place," such that "its road is the subject-matter of natural existence as science discovers and depicts it" (1925, p. 408).[5]

During the second half of the twentieth century, naturalistic philosophies rapidly gained support. As a result of this development, naturalism is contemporary philosophy's dominant "metaphilosophical ideology."[6] Leiter (2004, p. 3) even speaks about "a naturalistic turn" in contemporary philosophy. And although there has not been a large-scale reconstruction of this transition, it is safe to say that naturalism's contemporary prominence is for a significant part due to the work of Willard Van Orman Quine (1908–2000). If we take seriously the view that philosophy is not "an a priori propaedeutic or groundwork for science, but [. . .] continuous with science" (NK, 1969c, 126), Quine argues, we not only need to modify our beliefs about the supernatural, we also need to radically rethink our traditional *philosophical* views about truth, reality, justification, and meaning.[7] Quine's role in developing naturalism by applying it to traditional philosophical debates about truth, reality, justification, and meaning probably explains why he is often described as a "philosophers' philosopher" (e.g. in Føllesdal 2001).

[5] It should be noted that Dewey sometimes defines naturalism in weaker terms, for instance, when he claims that a naturalist is someone "who has respect for the conclusions of natural science" (1944, p. 2). Present-day naturalists too defend a wide range of metaphilosophical positions—some weak, some extremely strong. This situation has led many philosophers to complain that there are as many variants of naturalism as there are naturalists. See, for example, van Fraassen (1996, p. 172), who argues that it is "nigh-impossible" to "identify what naturalism is," but also earlier Seth (1896), Sellars (1927), and Nagel (1954, p. 3).

[6] See Kim (2003). Kim's claim is backed up by Bourget and Chalmers's (2014) recent survey among 931 leading philosophers. In response to the question whether they would describe themselves as naturalists or nonnaturalists in metaphilosophy, 49.8% of the respondents answered naturalist, whereas only 25.9% of the respondents chose the nonnaturalist option. One-quarter (24.3%) of the respondents were agnostic, insufficiently familiar with the issue, or thought that the question was too unclear to answer.

[7] Quine's important role in the naturalistic turn seems to be generally recognized. See Leiter (2004, p. 2), Hacker (2006, p. 231), Glock (2003, pp. 23–30), and Macarthur (2008, p. 2).

Quine is connected to his naturalistic predecessors in that he seems to have borrowed the *term* "naturalism" from Dewey. Although Quine, as we shall see, already defended a thoroughly naturalistic philosophy in the 1950s, it is only in his 1968 John Dewey Lectures that he first uses the term "naturalism" to describe his position—lectures in which he acknowledges Dewey as an intellectual predecessor:

> Philosophically I am bound to Dewey by the naturalism that dominated his last three decades. With Dewey I hold that knowledge, mind, and meaning are part of the same world that they have to do with, and that they are to be studied in the same empirical spirit that animates natural science. There is no place for a prior philosophy. (OR, 1968a, 26)

1.2. Naturalism and Analytic Philosophy

Despite the fact that Quine borrowed the term "naturalism" from Dewey, it would be a mistake to view his position as a continuation of the latter's ideas. Quine had already developed a naturalistic approach in all but name when he was still largely unfamiliar with Dewey's work.[8] And when he *did* start to study Dewey's work in the early 1970s, he largely disagreed with his instrumentalist reading of inquiry, i.e. with his picture of science "as a conceptual shorthand for organizing observations." In particular, Quine questioned whether Dewey's instrumentalism is consistent with his own view that we are always talking "within our going system when we attribute truth" and that we "cannot talk otherwise" (PPE, 1975b, 33–34).[9] Something similar can be said about the naturalistic turn in general. Despite the fact that naturalism has a rich history, ranging from nineteenth-century views about man, mind, and morality to the development of pragmatic naturalism in early twentieth-century American philosophy, it would be a mistake to view the naturalistic revolution in the second half of the twentieth century as a mere continuation of the developments sketched above.

To get a historically more accurate picture of both Quine's position and the naturalistic turn in general, they should be discussed against the background of developments within *analytic* philosophy. For not only did Quine develop his

[8] In fact, even after his John Dewey Lectures, Quine admitted that he was "not much of a Dewey scholar" (Quine to Pearl L. Weber, September 13, 1968, item DVFM 43/16208, Hickman 2008). See also Quine (IWVQ, 1994b, 70): "I was not influenced by Dewey. I didn't know his work that well in the old days." In section 7.5, I will reconstruct the events leading up to Quine's decision to adopt the term "naturalism" in the late 1960s.

[9] Quine's arguments against Dewey (and other pragmatists) are discussed in more detail in sections 4.2–4.3.

views largely in response to analytic philosophers like Bertrand Russell and Rudolf Carnap, the naturalistic turn too is generally viewed as a revolution that took place within analytic philosophy.[10]

Since its inception in the late nineteenth century, analytic philosophers have evinced a deep respect for the achievements of both the empirical and the formal sciences.[11] Yet, during the first half of the twentieth century, many analytic philosophers combined this respect for science with a firm antipsychologism, i.e. with the idea that psychological facts are irrelevant to questions of logical truth, justification, and meaning. Analytic philosophers, though generally applauding the developments in experimental psychology, believed that the field was categorically irrelevant to their inquiries into the question as to how we *ought* to reason.[12]

Mostly due to the influence of the early Wittgenstein, analytic philosophy's early antipsychologism evolved into a general distinction between science and philosophy, the former an empirical discipline concerned with fact, the latter an a priori discipline concerned with meaning. Many logical positivists and ordinary language philosophers eagerly adopted the distinction, dismissing the relevance of science for their inquiries.[13] In fact, the idea that there is a sharp distinction between science and philosophy was so pervasive among early analytic philosophers that Hacker (1996) has claimed that this distinction defines the tradition. Indeed, Hacker draws the radical conclusion that Quine's ideas

[10] See Kitcher (1992), Kim (2003, p. 84), and Leiter (2004, pp. 1–8). A notable exception is Hacker (2006, p. 231), who *does* see naturalism as a "pragmatist tradition." This, no doubt, has to do with his unusual definition of analytic philosophy, as will become clear below.

[11] Some even argue that analytic philosophy can be defined in terms of its respect for science. See Rorty (1982, p. 220), Wang (1986, p. 75), and Quinton (1995, p. 30) as quoted in Glock (2008a, p. 160).

[12] Most famous in this respect is Frege's attack on Mill's (1865, p. 359) claim that logic is a branch of psychology. Frege held that the "psychological is to be sharply separated from the logical" since logical truths are true regardless of whether we judge them to be true (1884, p. 17). In a similar fashion, the early Moore and Russell attacked idealism for its implicit reliance on psychological notions. They developed a metaphysics in which there is "no overt concern at all with the nature of thought or the mind or experience," as these notions were "looked on as psychological, and for this reason of no interest to philosophy" (Hylton, 1990, p. 108). For a wide-ranging study of the history of the debate on psychologism, see Kusch (1995).

[13] See Ricketts (1985), Coffa (1991), and Hacker (1996). Of course, there were exceptions (e.g. Otto Neurath). In fact, even Carnap's reconception of philosophy as the "logic of science" (Carnap, 1934, §72), often interpreted as supporting a paradigmatic science-philosophy distinction (i.e. the distinction between science and the logic of science), can be read naturalistically. Indeed, Carnap argues that the logic of science "is in the process of cutting itself loose from philosophy and becoming a properly scientific field, where all work is done according to strict scientific methods and not by means of 'higher' or 'deeper' insights" (Carnap, 1934, p. 46). See also Ebbs (2017) and Verhaegh (2017e).

"contributed [. . .] to the decline of analytic philosophy" (1996, p. xi) and that "[c]ontemporary philosophers who follow Quine have, in this sense, abandoned analytic philosophy" (1997, p. 10).

It is the background of this widespread distinction between science and philosophy that explains the revolutionary character of the naturalistic turn. And it was Quine, in the late 1940s and early 1950s, who contributed most to the decline of this distinction, arguing that it is impossible to draw a strict distinction between matters of fact and matters of language—between the analytic and the synthetic. In his first published dismissal of this distinction, "Two Dogmas of Empiricism" (1951), Quine argues (1) that the analytic-synthetic distinction cannot be drawn in a satisfying way and (2) that we do not require any such distinction in the first place.[14] Moreover, Quine also constructed an alternative, distinctively naturalistic approach to analytic philosophy; in *Word and Object* (1960), arguably the locus classicus of contemporary naturalism, Quine presents a metaphilosophy in which all forms of inquiry—philosophy, science, and common sense—are viewed as part of a single continuous enterprise.[15]

1.3. Working from Within

Despite Quine's substantial role in the contemporary naturalistic turn, many questions about his view remain to be answered. What are Quine's arguments for naturalism? What does it mean to say that philosophers are participants in the scientific enterprise at large? Why should we accept Quine's claim that there is no first philosophy? How should we interpret his view that philosophy and science are "continuous"? And how is his naturalism connected to his views about truth, reality, justification, and meaning? Although most contemporary Quine scholars agree that naturalism lies at the heart of his philosophy,[16] as of yet no

[14] For reasons that will become clear in the chapters to come, Quine was not very satisfied with "Two Dogmas." He believed that the paper did not contribute "a new idea to philosophy" (May 26, 1951, item 3015, my transcription) and that it had received "disproportionate attention" (June 18, 1951, item 1200, my transcription). Still, "Two Dogmas" deserves its status as a classic paper in the history of analytic philosophy because it remains his first *published* dismissal of the analytic-synthetic distinction.

[15] In focusing on *Word and Object*, I do not want to ignore the influence of "Epistemology Naturalized" (1969) on the naturalistic turn. Indeed, chapter 2 will be largely concerned with Quine's argument in that paper. As an aside, it is a little noted fact that Quine originally titled his paper "Epistemology Naturalized; or, the Case for Psychologism" (March 1968, item 2441), something which provides further support for the claim that his ideas ought to be viewed in the light of the antipsychologistic developments in early analytic philosophy.

[16] See, for example, Gibson (1988, p. 1), Hylton (2007, p. 2), Goldfarb (2008), Kemp (2012, p. 15), and Ebbs (2017).

one has systematically studied the nature and the development of Quine's naturalism at book length.

Unfortunately, Quine himself never systematically answered the above questions. Apart from his public lecture "Epistemology Naturalized" and a brief one-page discussion about "the sources of naturalism" in "Five Milestones of Empiricism,"[17] Quine never attempted to methodically discuss the nature of his most fundamental commitment. In fact, as we have seen, Quine does not even label his view "naturalistic" before the late 1960s. In the first forty years of his philosophical career, his naturalistic commitments were largely implicit.

This book makes amends by offering a systematic and a historical study of Quine's naturalism. I provide a detailed reconstruction of Quine's development, a novel interpretation of his arguments, and a systematic investigation into the arguments, theories, and commitments which led him to adopt the position. In the first part, I will use Quine's scattered remarks about naturalism and the traditional approaches to epistemology and metaphysics to develop a systematic account of Quinean naturalism. I argue that at the most fundamental level, to be a Quinean naturalist is to "work from within." Quine rejects the philosophical presupposition that it is possible to adopt a transcendental extrascientific perspective and replaces it with a rigorously science-immanent approach. I reconstruct Quine's arguments for such a science-immanent approach and argue that these arguments are based on his view that, as inquirers, we all have to start in the middle—in his view that we can only improve, clarify, and understand the system "from within" (FME, 1975a, 72).

In the second part, I turn to Quine's development. Just as Quine himself aims to understand the justificatory structure of science by adopting a genetic approach,[18] I examine the structure of Quine's most fundamental commitment by studying the evolution of his metaphilosophical thinking. Building on the wealth of material available at the W. V. Quine Papers—notes, academic and editorial correspondence, drafts, lectures, teaching materials, and grant proposals—I piece together the development and reception of Quine's epistemology, metaphysics, and metaphilosophy between 1927, when he first wrote about man's "web of knowledge" in a student paper (March 19, 1927, item 3225) and 1968, when he first decided to label his view "naturalistic." In the appendix, I have collected the papers, notes, and letters that have most affected my views about Quine's development.

[17] In chapter 2, I will discuss "Epistemology Naturalized." Quine's two "sources of naturalism" are examined in section 4.2.

[18] See, for example, (NNK, 1975d, 74–75): "We see [. . .] a strategy for investigating the relation of evidential support. [. . .] We can adopt a genetic approach, studying how theoretical language is learned. For the evidential relation is virtually enacted, it would seem, in the learning."

1.4. Reading Quine in Historical Context

Only thirty years ago analytic philosophy's self-image was still predominantly ahistorical. Analytic philosophers thought of themselves primarily as seekers of truth, not of historical understanding. That is, they were largely committed to the idea of an eternal, theoretically neutral framework for philosophical inquiry—unaffected by context, free of presupposition.[19]

To be sure, analytic philosophers did not neglect their intellectual predecessors: many analytic philosophers have written about a wide range of historical figures. Still, in engaging with their philosophical forerunners, they primarily treated them as contemporaries, ignoring conceptual and contextual differences. Analytic philosophers, in other words, have been prone to rationally reconstruct their predecessors' views in contemporary terms, such that their theories could be evaluated in terms of "philosophical truth and falsehood."[20]

Analytic philosophy's lack of concern with the historical mode of understanding is not surprising. For there seems to be a direct connection between these views on historiography and the antipsychologism discussed above. Since many analytic philosophers were interested only in logical, not in psychological, relations among propositions in their ordinary work, it should not be very surprising that they were also inclined to dismiss the relevance of both psychological and historical contexts in writing about their predecessors.[21]

Today the situation is entirely different. In the late 1980s and early 1990s, analytic philosophy witnessed what some have called a "historical turn." History of analytic philosophy is now widely viewed as an important field of study,

[19] See the introduction to Sluga's book on Frege: "From its very beginning, the [analytic] tradition has been oriented toward an abstracted, formal account of language and meaning, and not toward the comprehending of concrete historical processes" (1980, p. 2); as well as the preface to Hylton's (1990, p. vii) book on Russell: "Analytic philosophy has largely rejected historical modes of understanding. [. . .] [A]nalytic philosophy seems to think of itself as taking place within a single timeless moment." For a somewhat more distanced assessment of analytic philosophy's ahistoricism, see Peijnenburg (2000), Glock (2008b), and Reck (2013b).

[20] The phrase is Russell's. In the introduction to his book on Leibniz (1900, p. xx), Russell argues that "[p]hilosophical truth and falsehood [. . .] rather than historical fact, are what primarily demand our attention." Other often-mentioned examples of ahistoricist works in the history of analytic philosophy are Strawson's (1966) reconstruction of Kant, Bennett's (1971) work on Locke, Berkeley, and Hume, and Kripke's (1982) interpretation of the later Wittgenstein. See, for example, Watson (1993) and Beaney (2013b, p. 59).

[21] This link between antipsychologism and ahistoricism is explicitly made in Beaney (2013b, p. 37).

comprising a relatively large community of researchers, as is evinced by the ever-increasing stream of new monographs, papers, and collections.[22]

Whether there is a connection between the historical and the naturalistic turns is a matter of speculation and makes up an interesting subject for a different study.[23] Quine, in any case, did *not* combine his naturalism with a strong historical sensitivity. Although Quine includes history in his notion of "science" (RTE, 1997, 255), thereby suggesting that philosophy is also continuous with history, his more historically inclined work about his intellectual predecessors has been heavily criticized.[24]

As time passes, so do the frontiers of history. Where historians of analytic philosophy in the 1980s and 1990s were mainly interested in the nineteenth and the first half of the twentieth centuries,[25] today there is a growing historical interest also in philosophers who, like Quine, were largely active after the Second World War. In Quine's case, in particular, such a historical approach is much needed. For there is a general consensus among contemporary scholars that, despite the large body of books and papers about his philosophy, his work has been little understood.[26] Not only is Quine's work difficult to interpret because of its comprehensive character, subtle shifts in his philosophy throughout his seventy-year (!) academic career also require that one read him in *historical* context.

It is precisely this that I aim to do in the present study. In examining Quine's naturalism, I focus both on how his views on metaphilosophy, language, metaphysics, and epistemology systematically hang together, and on how his choices, theories, and arguments are influenced by conceptual and historical contingencies. In doing so, I hope to have found a balance between historical and systematic concerns and thereby to contribute both to the historiography of

[22] For an overview of some of the work, see Floyd (2009) as well as the essays collected in Floyd and Shieh (2001), Reck (2013a), and Beaney (2013a).

[23] On the one hand, it seems that in rejecting the sharp distinction between the logical and the psychological, naturalists create philosophical space for a more historically informed philosophy. Yet on the other hand, the more philosophers come to think of their inquiries as scientific in the sense of "natural science," the more they will be inclined to dismiss the relevance of the work of their predecessors. For some diverging opinions on the issue, see Taylor (1984), Rorty (1984), Hylton (1990, pp. 2–7), and, more recently, Williamson (2014, §3).

[24] See, for example, Friedman (1987, 1992) and Richardson (1990, 1992, 1998), who strongly dismiss Quine's interpretation of Carnap's (1928a) *Aufbau*. Notorious, in this respect, is Quine's quip that there are two sorts of people interested in philosophy: "those interested in philosophy and those interested in the history of philosophy." See MacIntyre (1984, pp. 39–40).

[25] Indeed, Floyd, in her overview, speaks about "the history of *early* analytic philosophy" as a field of study (2009, p. 157, my emphasis).

[26] See Kemp (2006, p. ix), Gregory (2008, p. 1), and Becker (2013, p. ix). Also with respect to naturalism in general, little historical work has been done, as has also been noted, in a somewhat different context, by Richardson (1997).

analytic philosophy and to a better, historically informed understanding of what is philosophically at stake in the contemporary naturalistic turn.

Finally, a general methodological disclaimer. Because I will be reading Quine in historical context, I will often be discussing his views about other philosophers, especially Russell and Carnap. As it is my primary aim to offer both a historical and a systematic reconstruction of *Quine's* position, I will mainly limit myself to *his* interpretation of these historical figures. I acknowledge, however, that Quine's reading does not always do justice to the historical Russell and Carnap.

1.5. Plan

This book is structured as follows. The first part offers an analysis of Quine's arguments against traditional epistemology (chapter 2) and traditional metaphysics (chapter 3) as well as a positive characterization of his naturalism (chapter 4). Together, these chapters provide a systematic overview of the nature of Quine's naturalism. The second part offers a historical reconstruction of Quine's development. I argue that Quine's development can be divided into three mutually reinforcing (and largely parallel) subdevelopments: (1) the evolution of his views about traditional metaphysics and epistemology (chapter 5); (2) his growing discontent with the analytic-synthetic distinction and his adoption of a holistic explanation of logical and mathematical knowledge (chapter 6); and (3) his growing recognition that his changing views about epistemology, metaphysics, and the analytic-synthetic distinction imply that science and philosophy are continuous (chapter 7).

In chapter 2, I start out with an analysis of Quine's argument in his most famous epistemological paper, "Epistemology Naturalized" (1969). On the basis of the argument in this paper, Quine's rejection of traditional epistemology (or "first philosophy") is often claimed to be based on an argument from despair. According to this standard conception, Quine rejects first philosophy because all attempts to reconstruct our scientific theories in terms of sense experience have failed. I show that this picture is historically inaccurate and that Quine's argument against traditional epistemology is considerably stronger than this received view suggests. According to Quine, the first philosopher's quest for foundations is useless; there is no science-independent perspective from which to validate science. I argue that Quine's stronger argument relies on a rejection of transcendental perspectives, and that a great deal of the confusion surrounding Quine's argument is prompted by certain phrases in "Epistemology Naturalized." Scrutinizing Quine's work both before and after the latter paper provides a better key to understanding his naturalistic views about the epistemological relation between theory and evidence.

With Quine's argument against traditional epistemology in focus, chapter 3 turns to Quine's position vis-à-vis traditional metaphysics. Prima facie, Quine's attitude toward metaphysics seems to differ from his attitude toward epistemology. For it is often claimed that Quine *saves* rather than dismisses traditional metaphysics in arguing that ontological questions are "on a par with questions of natural science" (CVO, 1951b, 211). Where Carnap rejects metaphysical existence claims as meaningless, Quine is often taken to restore their intelligibility by dismantling the former's internal-external distinction. The problem with this picture, however, is that it does not sit well with the fact that Quine, on many occasions, argues that metaphysical existence claims ought to be dismissed. Setting aside the hypothesis that Quine's metaphysical position is incoherent, I conclude that his views on metaphysics are subtler than is often presupposed; both the received view that Quine saved metaphysics *and* the opposite view that Carnap and Quine are on the same antimetaphysical team are too one-sided if we take seriously Quine's own pronouncements on the issue. Scrutinizing his published and unpublished work, I show that Quine is able both to blur the boundary between scientific sense and metaphysical nonsense *and* to argue that we cannot ask what reality is really like in a distinctively philosophical way—an argument, moreover, which shows that Quine's arguments against traditional epistemology and metaphysics are cut from the same cloth.

Having shown why Quine *rejects* traditional epistemology and metaphysics, I provide a more positive characterization of his naturalism in chapter 4. I argue that at the most fundamental level, to be a Quinean naturalist is to "work from within." I define Quine's naturalism in terms of two components—the principled rejection of transcendental perspectives and the adoption of a perspective immanent to our scientific conceptual scheme—and investigate what arguments, theories, and commitments have led him to adopt these theses. After examining Quine's own defense of naturalism in "Five Milestones of Empiricism," I argue that both components—no transcendence, and scientific immanence—as well as his deflationary theories of truth and justification, are based on the view that, as inquirers, we all have to start in the middle—in the view that we are always working from within.

Although Quine was always a science-minded philosopher, he did not adopt a fully naturalistic perspective until the mid-1950s. In the second part of this book, I reconstruct the genesis, the development, and the reception of Quine's naturalism. In chapter 5, I focus on the evolution of Quine's epistemology and metaphysics in the first twenty years of his career. Whereas his published work in the 1930s and 1940s was primarily technical, the Quine archives reveal that he was already working on a philosophical book during the Second World War—a project entitled *Sign and Object*. I argue that *Sign and Object* sheds new light on the evolution of Quine's naturalism. Where

"Two Dogmas of Empiricism" is usually considered to be a turning point in Quine's development, this chapter redefines the place of "Two Dogmas" in his oeuvre. Not only does Quine's book project show that his views were already fairly naturalistic in the early 1940s, *Sign and Object* also unearths the steps Quine had to take in maturing his perspective.

Having reconstructed Quine's development in epistemology and metaphysics, in chapter 6 I turn to his evolving views on the analytic-synthetic distinction. Following his two-tiered argument in "Two Dogmas of Empiricism," I answer three questions: (1) when did Quine start demanding a *behavioristically acceptable* definition of analyticity? (2) when did he stop searching for such a definition? and (3) when did he reject the dogma of reductionism, concluding that there is no need for an analytic-synthetic distinction in the first place? I argue that it is impossible to identify a specific moment at which Quine definitively rejected the analytic-synthetic distinction; from a developmental perspective, all three questions require an independent answer. In the second part of chapter 6, I reconstruct Quine's evolving views on the analytic-synthetic distinction *after* 1951 and challenge the common misconception that he changed his mind about logic, holism, and analyticity in the later stages of his career.

In chapter 7, finally, I turn to Quine's evolving views on the relation between science and philosophy. Scrutinizing both the development of and the external responses to *Word and Object*, arguably the locus classicus of his naturalistic worldview, I examine how Quine became increasingly aware of the metaphilosophical implications of the views he had first developed in the late 1940s and early 1950s. In the final sections of this chapter, I offer a detailed historical reconstruction of Quine's decision to label his position "naturalism."

This book, in sum, offers both a historical and a systematic investigation of naturalism as it was developed by Quine. It could not have been written without the wonderful collection of documents at the W. V. Quine archives at Houghton Library. In the appendix, I have collected and transcribed the documents that have most affected my views about (the development of) Quine's naturalism. It contains two previously unpublished essays, two transcribed sets of handwritten notes, and a previously unpublished letter to Nelson Goodman.

PART I

NATURE

2

Naturalizing Epistemology

2.1. Introduction

According to Quine, naturalism can be characterized negatively as the abandonment of "the goal of a first philosophy" prior to science (FME, 1975a, 72). But what exactly are Quine's reasons for rejecting first philosophy? Why, in other words, does Quine believe that there is no "a priori propaedeutic or ground work for science" (NK, 1969c, 126)? In the present chapter, I examine Quine's ideas about first philosophy and reconstruct his argument for dismissing the project.

Prima facie, Quine's argument against first philosophy seems to be pretty straightforward: we ought to abandon traditional epistemology because, historically, all attempts to ground our beliefs have failed. In "Epistemology Naturalized" (1969a), Quine divides traditional epistemology into a doctrinal and a conceptual program and argues that neither project can be carried out satisfactorily. On the doctrinal side, Hume's problem of induction prevents us from deducing our beliefs about the world from basic observation statements. On the conceptual side, Quine criticizes the epistemologist's attempts to translate our theoretical concepts in sensory terms. In particular, he criticizes Carnap's project of rational reconstruction, arguing that it fails to "offer any key to translating the sentences of science into terms of observation, logic, and set theory" (EN, 1969a, 77). As an alternative to these projects, Quine proposes his naturalized epistemology, the study of how theory and evidence are *actually* related:

> If all we hope for is a reconstruction that links science to experience in explicit ways short of translation, then it would seem more sensible to settle for psychology. Better to discover how science is in fact developed and learned than to fabricate a fictitious structure to a similar effect. (p. 78)

Where the traditional epistemologist rejects naturalism as circular—we cannot use science in justifying science—Quine believes that he is free to use scientific knowledge in his inquiries: "scruples against circularity have little point once we have stopped dreaming of deducing science from observations" (p. 76).

Let me call this the *standard conception* of Quine's argument against first philosophy. According to the standard conception, we are justified in adopting a naturalized epistemology only after we have established that all attempts to reduce our knowledge to sense experience have failed. Quine's argument, in other words, is construed as a conditional argument: we can legitimately take on a naturalized epistemology only when we have demonstrated that we ought to "stop dreaming of deducing science from sense data" (EN, 1969a, 84) and that we ought to "despair of being able to define theoretical terms generally in terms of phenomena" (FME, 1975a, 72). The Quinean naturalist is not a "busy sailor" from birth, but "someone who later elects to enlist, perhaps in reaction to some deep disappointment" (Maddy, 2007, 85). Quine's argument, in short, is pictured as an *argument from despair*.[1]

The standard conception is widespread among both Quine scholars and critics. In "The Key to Interpreting Quine," for example, Roger F. Gibson summarizes Quine's arguments against the doctrinal and the conceptual program and concludes that "[t]he thesis that there is no first philosophy is a comment on the failure of traditional epistemologists to find a foundation outside of science upon which science [. . .] can be justified" (1992, p. 17). Similarly, P. M. S. Hacker claims that "[t]he failure of the Carnapian enterprise seemed to Quine to warrant the naturalization of epistemology" (2006, p. 236), and Penelope Maddy argues that the Quinean naturalist seems to be "driven to her position by 'despair' at the failure of any or all attempts to 'ground' science" (2007, p. 85).[2]

Still there seems to be something odd about the standard conception. For one thing, Quine's argument from despair occurs only in "Epistemology Naturalized" (1969a) and in "Five Milestones of Empiricism" (1975a). The argument is conspicuously absent in Quine's work before and after these two papers, even when he discusses the distinction between traditional and naturalized epistemology. This gap is particularly apparent in *From Stimulus to Science* (1995b). In the first chapter of this book, Quine gives an extended summary of the traditional quest for certainty, starting with skeptical worries about our knowledge of the external world and ending with Carnap's project of rational reconstruction. Yet in the

[1] This apposite phrase is David Shatz's: "Quine arrived at [his] proposal by route of an argument we might term the argument from despair. The traditional project of validating common sense and scientific beliefs in the face of skeptical challenge has been, and is doomed to be, a failure; therefore, the project is best dropped" (1994, p. 117). According to Shatz, the alternative to an argument from despair is a dialectical naturalism, which aims to "confront the problem of skepticism and of circularity head on." Shatz believes that Quine in some places "provides a partial defense of dialectical naturalism" (1994, p. 120).

[2] See also, for example, Roth (1999, §2), Kertész (2002, §3), and Fogelin (2004, pp. 19–27).

second chapter, which deals with his naturalism, Quine nowhere uses the traditional epistemologists' failed attempts to derive science from sense data as an argument for adopting a naturalistic perspective. Rather, he reflects on the "phenomenalistic orientation" of the traditional project, i.e. about "[t]he idea of a self-sufficient sensory language as a foundation for science" (FSS, 1995a, 15).

Second, if the argument from despair were all he had to offer, Quine would not have made a particularly strong case for the naturalization of epistemology. For as many epistemologists have objected, it is one thing to dismiss the traditional quest for absolute foundations; it is quite another thing to reject the search for justification tout court and to claim that "[e]pistemology, or something like it, simply falls into place as a chapter of psychology and hence of natural science" (EN, 1969a, 82).[3] Quine argues only that we cannot *completely* ground our beliefs on sense experience, an argument that is too weak to convince any first philosopher who shares Quine's doubts about our ability to answer the Cartesian skeptic. In response to Quine's despair, traditional epistemologists could easily adopt a "*moderate* first philosophy, which eschews certainty but which allows for the independence (of epistemology from science) sought by the traditionalist" (Siegel, 1995, p. 53).

In this chapter, I argue that the standard conception is mistaken. I show that Quine's argument against the first philosopher is considerably stronger than the received view suggests. In works both before and after "Epistemology Naturalized," Quine does not abandon traditional epistemology out of despair but because the project is demonstrably flawed from the beginning. According to Quine, it is a mistake to believe that one can develop a self-sufficient sensory-language, independent of our best scientific theories of the world. The first philosopher does not fail because he aims at Cartesian certainty, but because he presupposes that he can adopt some science-independent perspective. I argue, in short, that "Epistemology Naturalized," when considered in isolation from the rest of his work, misrepresents the strength of Quine's position.[4]

[3] The loci classici of this argument are Putnam (1982) and Kim (1988). To the best of my knowledge, Giedymin (1972) was the first to address this issue.

[4] A great deal of the confusion on the part of the standard conception seems to be triggered by certain phrases in "Epistemology Naturalized." I am not the first to point to the somewhat problematic relation between this paper and the rest of Quine's work. See Johnsen (2005). It should be noted in this respect that "Epistemology Naturalized" was written for a general public. In a report about his trip to Vienna, where he first presented "Epistemology Naturalized," Quine writes:

> On September 9, I gave a one-hour address, "Epistemology naturalized." It was one of several invited addresses arranged by the Congress for the Viennese public; in participating I represented the United States, by selection of the American Philosophical Association. It was in appreciation of this honor, and thus with quasi-diplomatic rather than scientific

What I offer, then, is a reconstruction of Quine's actual argument against first philosophy, focusing on his work both before and after "Epistemology Naturalized." This chapter is structured as follows. I start by outlining the standard conception and by examining Quine's argument from despair (section 2.2), after which I introduce his stronger argument (section 2.3) and show how he uses it to dismiss both the traditional epistemologist (section 2.4) and the skeptic (section 2.5). I end this chapter with an analysis of how we might better read "Epistemology Naturalized" in the light of these findings (section 2.6).

2.2. From Certainty to Straight Psychology

Although we seem to know a great many things about ourselves and the world around us, we can never be absolutely sure that our beliefs are true. Even our best scientific theories, history has taught us, might turn out to be false or to rest on misguided assumptions. According to Quine, traditional epistemology starts from a deep dissatisfaction with this situation: "the theory of knowledge has its origin in doubt, in scepticism. Doubt is what prompts us to try to develop a theory of knowledge" (NNK, 1975d, 257).[5] In order to restore confidence in both our everyday convictions and our scientific theories, traditional epistemologists seek to ground our beliefs upon something more secure. Quine often refers to this project as "the Cartesian dream," the dream of an indubitable foundation for our beliefs about ourselves and our surroundings.[6]

In analyzing the epistemologists' quest for certainty, Quine has focused almost exclusively on *empiricist* attempts to ground our knowledge.[7] In his discussion of empiricist epistemology, Quine distinguishes two projects, one

> motives, that I consented to go to the Congress. In respect of intellectual content I do not find such occasion very rewarding. (November 19, 1968, item 1239)

See also Quine's autobiography, where he describes "Epistemology Naturalized" as a "public address" (TML, 1985a, 350).

[5] See also (LDHP, 1946c, 50–51) and (FSS, 1995a, 1).

[6] See (PT, 1990a, 19). In referring to the quest for certainty as "the Cartesian dream," Quine is referring to Descartes's tremendous influence in putting the search for foundations on the philosophical agenda. Quine is aware, however, that Descartes was not the first to attempt to ground our knowledge about the world. He recognizes that "the quest for certainty goes back to Plato, and nobody knows how far beyond" (LDHP, 1946c, 52).

[7] An exception is (LDHP, 1946c, 54–59), where he explicitly discusses and rejects *rationalism*. Most notably, Quine questions whether the rationalist can be certain that her innate ideas are true, even if they seem self-evident. See section 4.2.

doctrinal and one conceptual. The doctrinal project is concerned with truth and aims at deriving our beliefs about the world, especially our well-established scientific theories, from basic observation statements. The conceptual project, on the other hand, is concerned with meaning and aims at translating our scientific concepts in sensory terms. The two projects are connected: if one succeeds in defining all scientific concepts in sensory terms, then one's scientific beliefs and one's basic observation statements will be couched in the same sensory language, an accomplishment that will enable one to examine whether the former can be derived from the latter (EN, 1969a, 69–71).

According to Quine, the classical empiricists failed in both respects. On the conceptual side of epistemology, Locke, Berkeley, and Hume were unable to indicate how our complex ideas about the world can be constructed out of simple ones; defining even the very notion of an enduring physical body turned out to be problematic. Still, their problems were even worse on the doctrinal side. For it was Hume who showed that it is impossible to establish a deductive relation between theory and evidence even if both are couched in the same sensory language; neither general statements nor singular statements about the future can be deduced from any finite set of sensory evidence (pp. 71–72).

Quine is convinced that there is no progress to be made with respect to the doctrinal project: "The Humean predicament is the human predicament" (p. 72). Although the value of inductive reasoning in science can hardly be overestimated, the traditional epistemologist simply has to admit that we are never strictly entitled to rely on induction (RA, 1994d, 231–33). Still there was progress to be made with respect to the conceptual project. Quine argues that some major advances in the eighteenth and nineteenth centuries breathed new life into the empiricists' program.

One of these developments was Jeremy Bentham's work on contextual definition, or what he called *paraphrasis*. The classical empiricists had developed what Quine has called a "term-by-term empiricism" (TDE, 1951a, 42).[8] Their goal was to define our complete vocabulary in sensory terms. Bentham showed,

[8] Strictly speaking, classical empiricism is better characterized as an "idea-by-idea" empiricism, since its aim is to define complex *ideas* in terms of simple ones. Quine often credits John Horne Tooke, a contemporary of Hume, for shifting the empiricists' attention from ideas to words: "Tooke appreciated that the *idea* idea itself measures up poorly to empiricist standards" (FME, 1975a, 68). See also (TDE, 1951a, 38–39), (FM, 1977, 271–72), and (FSS, 1995a, 6). According to Quine, the "idea idea" was a leftover from Cartesian rationalism. Where rationalists stressed that we should ground our beliefs upon clear and distinct ideas, the classical empiricists rejected innate ideas and replaced them with the impressions we obtain through the use of our senses. Still, even though they disagreed with the rationalists on the *source* of our ideas, the empiricists maintained the view that our beliefs about the world should be grounded in *ideas* that are clear and distinct (LDHP, 1946c, 63).

however, that terms can also be defined contextually; one can define a term simply by showing how all sentences containing the term can be paraphrased into sentences without it. In general, one can contextually define a word *W* on the basis of some accepted defining vocabulary *V*, by explaining how to paraphrase every sentence *S* in which *W* occurs into a new sentence that contains only words of *V* and *S* other than *W* (VD, 1972, 55).[9]

According to Quine, Bentham's method proved to be greatly beneficial for the empiricists' conceptual studies. Instead of defining our complete theoretical vocabulary in sensory terms, the empiricist could now also rely on contextual definition:

> Hume's [. . .] desperate measure of identifying bodies with impressions ceased to be the only conceivable way of making sense of talk of bodies, even granted that impressions were the only reality. One could undertake to explain the talk of bodies in terms of talk of impressions by translating one's whole sentences about bodies into whole sentences of impressions, without equating the bodies themselves to anything at all. (EN, 1969a, 72)

Next to Bentham's method of paraphrasis, Quine lists the development of set theory in the late nineteenth century as an advancement that led to substantial progress on the conceptual side of the empiricists' quest for certainty. Taking our sense impressions as the fundamental objects of a set theoretic structure, the empiricist "is suddenly rich: he has not just his impressions to play with, but sets of them, and sets of sets, and so on up" (p. 73); he has access to an infinite universe of sets containing all possible combinations of impressions.

According to Quine, Russell was the first the see the epistemological potential of these logical and mathematical advancements.[10] Still, Quine credits Carnap as the philosopher who actually attempted to carry out the project by using these formal tools to construct our beliefs about the world out of primary sense experiences. According to Quine, Carnap's *Der Logische Aufbau der Welt* (1928a) constituted "a masterful construction" of the external world from the

[9] See also (CD, 1995b). As an example, consider the mathematical expression $\frac{\sin \pi}{\cos \pi}$. This expression may be abbreviated as $\tan \pi$ without defining "tan" directly in terms of "sin" and "cos." It is enough to relate definiendum and definiens contextually such that for all x, $\tan x$ equals $\frac{\sin x}{\cos x}$ (TC, 1936, 78).

[10] See (ROD, 1966c, 83–84) and (EN, 1969b, 73–74) for Quine's reading of Russell's *Our Knowledge of the External World* (1914). It should be noted, however, that Quine's reading is not uncontroversial. According to Pincock (2007), Nasim (2012), and Klein (2017), for example, Russell should not be interpreted as reviving the empiricist project.

data of sensation "using the sophisticated devices of mathematical logic" (CA, 1987b, 144).[11]

Carnap opens the *Aufbau* with the claim that he attempts to establish a *constructional system*, "a step-by-step derivation or 'construction' of all concepts from certain fundamental concepts" (Carnap, 1928a, §1). The concept he chooses as the foundation of his construction is what he calls an *elementary experience*, an individual's totality of experiences at a given moment in time involving all sense modalities.[12] Next to a fundamental concept, Carnap also introduces the dyadic predicate Rs. This relation Rs holds between two elementary experiences *x* and *y* whenever the subject recognizes *x* and *y* as partially similar (§78). Using only these very basic elements,[13] Carnap manages to define "a wide array of important additional sensory concepts which, but for his constructions, one would

[11] Here too it should be noted that this is *Quine's* interpretation of the *Aufbau*. Friedman (1987, 1992), Tennant (1994), and Richardson (1990, 1992, 1998), among others, criticize Quine's reading and have developed an interpretation in which Carnap's choice to work from a phenomenalist basis in the *Aufbau* was merely arbitrary. In this reading, Carnap's intention "is not so much to give a traditional empiricist justification for our knowledge of the external world as to exhibit what Carnap calls the 'neutral basis' common to *all* epistemological views—whether empiricist, transcendental idealist, realist, or subjective idealist" (Friedman, 2007, p. 5). In a reply to Tennant, Quine recognizes that there are some passages in the *Aufbau* that support such a neutralist reading, but he maintains that the *Aufbau* was initially supposed to be part of the phenomenalist project:

> I have a further hypothesis. [. . .] I picture Carnap as having been a single-minded phenomenalist when he devised the constructions that went into the *Aufbau*. When the book was about ready for printing, I picture Neurath pressing the claims of physicalism. I then picture Carnap writing and inserting those paragraphs of disavowal by way of reconciling the book with his changing views. Significantly, he took the physicalistic line in his subsequent writings, and refused permission to translate the *Aufbau* for more than thirty years. (CNT, 1994a, 216)

Interestingly, Quine already interpreted the *Aufbau* as a phenomenalistic work when he first discussed the book with Carnap during his trip to Europe in 1933. In a letter to John Cooley, Quine explains: "The *Aufbau* stands to [the] philosophical doctrines of Russell and Lewis as *Principia* stands to the antecedent purely philosophical suggestion that mathematics is a form of logic. [. . .] Carnap has since departed [. . .] from the point of view of the *Aufbau*" (April 4, 1933, item 260, my transcription). Whether or not this reading is correct, in what follows I mainly limit my discussion to Carnap's project as Quine conceives it. See section 1.4.

[12] Note that this description of an elementary experience only makes sense from the point of view of the finished construction; individuals and sense modalities are not presupposed at the start of Carnap's project.

[13] In fact, Carnap succeeds in defining elementary experiences in terms of Rs, such that his construction makes essential use only of logic, mathematics, and one dyadic predicate. Moreover, in sections 153–55, Carnap attempts to eliminate Rs as well by introducing the notion of *foundedness* into his logic and by defining Rs in terms of it. Scholars have been critical of Carnap's success on this score, however. See, for example, Friedman (1987, pp. 532–33).

not have dreamed were definable on so a slender basis" (TDE, 1951a, 39). Most importantly, Carnap succeeds in constructing the five sense modalities as well as the basic sense qualities that are taken for granted in the classical empiricists' epistemological framework.

After having constructed these basic sensory concepts, Carnap attempts to step outside the subjective arena of experience into the intersubjective world. As a first step, Carnap wants to assign the sense qualities in our visual field—i.e. the colors in our two-dimensional visual space—to points in the three-dimensional physical space order, a maneuver that Carnap believes to be "one of the most important steps in the constructional system" (Carnap, 1928a, §124). The idea, as Quine notes, is to translate sentences of the form 'Quality q is at point-instant x ; y; z; t' in terms of the fundamental notions that Carnap allows in his constructional scheme (TDE, 1951a, 40).

As he himself has recognized in the preface to the second edition of the *Aufbau*, however, Carnap did not succeed in constructing the intersubjective "is at" connective from any of the subjective lower-level concepts:

> One of the most important changes [in the second edition] is the realization that the reduction of higher level concepts to lower level ones cannot always take the form of explicit definitions. [. . .] Actually, without clearly realizing it, I already went beyond the limits of explicit definitions in the construction of the physical world. For example, for the correlation of colors with space-time points, only general principles, but no clear operating rules were given. (1928a, p. viii)

Instead of providing a full translation of our color-assignments, Carnap was able to provide only a list of desiderata that any assignment of colors to space-time points should satisfy "as far as possible," while being aware that they can never be "precisely satisfied" (§126).

It is important to see why Carnap's reduction broke down at this point. Carnap's desiderata for assigning colors to world points prescribe only a complete assignment, not a point-by-point allocation. The reason for this is that one needs to distinguish between genuine information from the outside world and subjective color experiences such as hallucinations and disturbances of the eye (§126). The problem for Carnap is that one can only judge whether some experience is hallucinated when one examines whether it fits in one's total allocation of visual experiences over time. One cannot judge whether a single experience is hallucinated on the basis of that very experience alone; "the assignment of sense qualities to public place-times has to be kept open to revision in the light of later experience, and so cannot be reduced to definition" (ROD, 1966a, 85). In consequence, one cannot assign one color to a particular space-time point without considering its place holistically in the total color-to-world allocation. Carnap's construction broke down, in other words, because he failed to take into account the holistic nature of the theory-evidence relation.[14]

In response to his failure to develop a satisfactory criterion of empirical significance, Carnap radically altered his views after the *Aufbau*. In "Testability and Meaning" (1936, 1937), Carnap gave up on the idea that theoretical sentences should be strictly translatable into the observation language if they are to be empirically significant. Instead, he introduced a liberal form of reduction that allows theoretical sentences to be correlated with lower-level sensory sentences in a way short of translatability. Rather than demanding strict reductions such that theoretical sentences are eliminated in favor of observation sentences, Carnap now also admitted *reduction sentences* that define new theoretical terms relative to specified experimental conditions.[15]

In "Epistemology Naturalized," Quine seems to argue that Carnap's adjustments were fatal for traditional epistemology. For in dispensing with reduction by elimination, "the empiricist is conceding that the empirical meanings of typical statements about the external world are inaccessible and ineffable"

[14] See also (CPT, 1984a, 125–26): "A typical single sentence of a theory has no distinctive empirical content of its own; it can be singled out for testing, but only by agreeing meanwhile to hold other sentences of the cluster immune [. . .] in the *Aufbau* the very mechanism of [this] Duhem effect is strikingly and imaginatively depicted."

[15] Carnap (1936, §8) defines reduction sentences as follows. Let Q_3 be a theoretical predicate, let Q_1 and Q_4 describe experimental conditions which have to obtain in order to find out whether or not a space-time point *b* has the property Q_3, and let Q_2 and Q_5 describe possible results of the experiments. Then Q_3 can be introduced as a new predicate in one's language by means of the following pair of sentences R_1 and R_2:

$$(R_1)\ Q_1 \rightarrow (Q_2 \rightarrow Q_3)$$
$$(R_2)\ Q_4 \rightarrow (Q_5 \rightarrow \neg Q_3)$$

Definitions of this form are *partial* definitions because the meaning of Q_3 is only specified relative to a set of experimental conditions Q_1 and Q_4.

Shortly after "Testability and Meaning" (1936, 1937) Carnap recognized that even these partial definitions are not yet liberal enough; sentences containing highly theoretical concepts like "absolute temperature" and "wave function" resist such an interpretation. Again the problem was the holistic character of the theory-evidence relation. In the laboratory, a negative test result does not necessarily imply that a certain disposition is not present (the scientist can always maintain her belief that a disposition is present by revising one of her auxiliary hypotheses). Similarly, a positive test does not imply that the disposition *is* present. Reduction sentences, however, do not allow for these possibilities, as they are intended to state necessary conditions for the application of a term. See Hempel (1952, p. 32) and Carnap (1956, p. 68). In response, Carnap proposed an even more liberal criterion of empirical significance, one which recognizes that "[t]he definition of meaningfulness must be relative to a theory *T*, because the same term may be meaningful with respect to one theory but meaningless with respect to another" (1956, p. 48). Yet even this definition did not fully implement the lessons of holism, as Quine shows (CPT, 1984a, 125).

(EN, 1969a, 78–79). That is, in allowing more liberal forms of reduction, Carnap acknowledged that he would never be able to completely specify the empirical meanings of isolated theoretical sentences. "Epistemology Naturalized," in other words, seems to construe Carnap's concession as a natural endpoint for traditional epistemology. Where Hume had already demonstrated that we cannot hope to fulfill the doctrinal project, Carnap's *Aufbau* showed that the conceptual project is too demanding as well. Quine argues that to "relax the demand for definition, and settle for a kind of reduction that does not eliminate, is to renounce the last remaining advantage that [. . .] rational reconstruction [had] over straight psychology; namely, the advantage of translational reduction" (EN, 1969a, 78). We ought to "stop dreaming of deducing science from sense data" (p. 84) and we ought to "despair of being able to define theoretical terms generally in terms of phenomena" (FME, 1975a, 72). Hence, we are better off studying the actual relation between theory and evidence.

2.3. Two Strategies

The argument outlined above is mainly concerned with the empiricists' ideas about the relation between theory and evidence, with their attempts to connect our scientific beliefs with our primary sense experiences. Quine argues that we are unable to ground our beliefs on sense experience and that we cannot translate our theoretical vocabulary in observational terms. Schematically, the problem is that we have (A) our primary sense-experiences, and (B) our best scientific theories, but that we do not seem to be able to relate (A) and (B) in an epistemologically satisfying way. The holistic character of the theory-evidence relation prevents us from establishing an epistemologically satisfying connection between the two because a typical single (B)-sentence "has no distinctive empirical content of its own" (CPT, 1984a, 125).[16]

Still, criticizing the epistemologist's ideas about the relation between theory and evidence is not the only way to challenge the traditional project. There remains a second option. Instead of showing that all attempts to base

[16] The empirical content of a theory is the set of observation categoricals that it implies. Observation categoricals are sentences of the form "Whenever P, Q," where P and Q are observation sentences such that the categorical expresses "the general expectation that whenever the one observation sentence holds, the other will be fulfilled as well." As examples of observation categoricals, Quine mentions "When it snows, it's cold," "Where there's smoke, there's fire," and "When the sun rises, the birds sing" (FSS, 1995b, 25). Quine's claim that a typical (B)-sentence has no distinctive empirical content of its own, in other words, implies that one cannot derive an observation categorical from a single theoretical sentence; only clusters of theoretical sentences imply observation categoricals. I will examine Quine's holism in more detail in sections 4.2 and 6.3.

our scientific beliefs on some science-independent foundation have failed, one can also attempt to criticize the very idea of a science-independent foundation itself. That is, instead of challenging the *nature* of the relation between (A) and (B), one can also call into question the *epistemological value* of connecting (B) with (A) in the first place. One could, for example, dismiss the traditionalist's ideas about the epistemological status of (A) and argue that sense experience does not constitute a truly science-independent foundation to begin with.

In "Epistemology Naturalized," Quine does not discuss this second option.[17] That is, he does not question the idea of a self-sufficient sensory language presupposed in the epistemologists' attempts to reduce science to sense experience. Quine argues only that *once* we have adopted a naturalized epistemology, we can substitute our talk about sense data with talk about its scientific analogue, the physical stimulation of our sensory receptors:

> [O]ne *effect* of seeing epistemology in a psychological setting is that it resolves a stubborn old enigma of epistemological priority. [. . .] In the old epistemological context [. . .] we were out to justify our knowledge of the external world by rational reconstruction, and that demands awareness. Awareness ceased to be demanded when we gave up trying to justify our knowledge of the external world by rational reconstruction. What to count as observation now can be settled in terms of the stimulation of sensory receptors. (EN, 1969a, 84, my emphasis)

In the remainder of this chapter, I argue that both before and after "Epistemology Naturalized," Quine argues exactly the other way around. Quine does not give up on sense data because of his naturalism. Rather, he naturalizes epistemology because of his doubts about the idea of "a self-sufficient and infallible lore of sense data" (NLOM, 1995c, 462). That is, Quine's doubts about "epistemological priority" are not a *consequence* of his naturalism; they are the very *reason* he adopts a naturalized epistemology in the first place. Both before and after "Epistemology Naturalized," I will show, Quine *does* use the second strategy; he criticizes the traditional project because he believes that attempts to connect (A) and (B) are futile from an epistemological perspective.

[17] At least, Quine does not discuss this second option when it concerns the empiricist program of reducing science to sense data. Quine *does* use the second strategy when he dismisses the logicist program of reducing mathematics to logic and set theory. Quine argues that the logicists failed because their foundations were not truly mathematics-independent. According to Quine, set theory is itself a branch of mathematics, and so the logicists failed to do "what the epistemologist would like of it," i.e. revealing the ground of mathematical knowledge (EN, 1969b, 70).

2.4. Self-Sufficient Sensory Languages

From the very beginning of his philosophical career, Quine thought about what he calls the problem of "epistemological priority," the problem of how to reconcile phenomenalistic and physicalistic conceptual schemes. Already in his one of his graduate papers, for example, Quine questioned the nature of the phenomenalist's view about immediate experience:

> [W]hat are those "bare data" [. . .]? Certainly they are themselves a high refinement of abstraction. My experient career is not a simple matter of consciously taking odds and ends and amorphous bits of unidentified data and fitting them into a system; what I see before me is a *chair*, not an array of varicolored quadrilaterals which I consciously assemble and classify as a chair. My immediate experience, rather than consisting of raw material to be interpreted, is already seething with interpretation. (March 19, 1931, item 3225)

Still, for quite some time, Quine left open the option that sense data are *epistemologically* prior, that the phenomenalistic conceptual scheme is fundamental from an epistemological point of view. In "On What There Is" (1948), for example, Quine argues that

> from among the various conceptual schemes best suited to [. . .] various pursuits, one—the phenomenalistic—claims epistemological priority. Viewed from within the phenomenalistic conceptual scheme, the ontologies of physical objects and mathematical objects are myths. The quality of myth, however, is relative; relative, in this case, to the epistemological point of view.[18] (OWTI, 1948, 19)

Quine's position was a pragmatic one: we want an ontology that is as simple as possible but both conceptual schemes are simple in their own respects. A phenomenalistic scheme is simple because it posits only subjective events of sensation, whereas a physicalistic scheme can be said to offer conceptual simplicity (OWTI, 1948, 17).

Despite this pragmatic attitude,[19] however, Quine was aware that we cannot reduce our complete vocabulary to sensory terms, i.e. that the idea of a complete rational reconstruction is an idle dream: "there is no likelihood that each

[18] Note that, in talking about "conceptual schemes" here, Quine is not invoking a distinction between conceptual schemes and languages; for Quine they are one and the same. See (VITD, 1981d, 41).

[19] As we shall see in chapter 5, it is not completely correct to think of Quine as defending a pragmatic position on this issue in the 1930s and 1940s. For our present purposes, however, it will do.

sentence about physical objects can actually be translated [. . .] into the phenomenalistic language" (OWTI, 1948, 18). A few years later, in "Two Dogmas Empiricism," Quine of course explained why such a strict reduction is impossible: "our statements about the external world face the tribunal of sense experience not individually but only as a corporate body" (TDE, 1951a, 41).

As a result, the main ingredients of "Epistemology Naturalized" were already in place in the early 1950s: Quine was already familiar with the possibility of adopting a purely physicalistic conceptual scheme, and he had already shown that the traditional epistemologists' attempts at reduction were fruitless. Still, Quine had not yet adopted a naturalized epistemology at this point. He still believed that there might be epistemological reasons for adopting a phenomenalistic conceptual scheme. This situation had not changed in "Two Dogmas," where Quine continued to talk about "sense data" in describing the evidential boundaries of his newly developed holistic empiricism. In a 1951 letter to James B. Conant, for example, Quine explains:

> The philosopher who epistemologizes backward to sense data [. . .] is fashioning a conceptual scheme just as the physicist does; but a different one, for a different subject. [. . .] He agrees that there are electrons and tables and chairs and other people, and that the electrons and other elementary particles are 'fundamental' in the physical sense. [. . .] But sense data are 'fundamental' in the epistemological sense. (April 30, 1951, item 254).[20]

Between "Two Dogmas" (1951a) and *Word and Object* (1960) however, Quine *did* switch exclusively to a physicalistic conceptual scheme. Looking back on this period, Quine has referred to the ten years between these two works as the decade in which he became "more consciously and explicitly naturalistic," as the period in which he "stiffened up his flabby reference to 'experience' by turning to our physical interface with the external world: the physical impacts of rays and molecules upon our sensory surfaces" (TDR, 1991b, 398). That is, in the decade following "Two Dogmas," Quine adopted a physicalistic conceptual scheme and started to talk exclusively about the stimulation of sensory receptors.[21]

[20] See also the first edition of *Methods of Logic*. In the introduction to this first edition, Quine claims that "[t]he seeing of a green patch, and the simultaneous utterance 'Green patch now', constitute the sort of composite event which, in its rare occurrences, gladdens the heart of the epistemologist" (ML1, 1950a, xi). See Murphey (2012, p. 89). Interestingly, Quine deleted this in the fourth edition of *Methods of Logic*. There he only talks about "the utterance of a statement on the occasion of a stimulation to which that string of words has become associated" (ML4, 1982, 1). I thank Thomas Ricketts for this suggestion.

[21] Again, a more detailed historical account of the evolution of Quine's naturalism in the first decades of his career will be provided in chapters 5 and 6. In this section, I focus solely on Quine's reasons for abandoning phenomenalism.

So why did Quine give up on phenomenalism? Did Quine give up on sense datum languages out of despair? Did he, in other words, come to regard the traditional perspective as hopeless because we cannot reduce science to sense experience? No, he did not. What changed is that he became convinced that the very idea of a sense datum language is not *epistemologically prior to* but *dependent on* our best scientific theories of the world—that "[s]ense data are posits too" (PR, 1955, 252). Quine came to believe that the traditional project was flawed from the beginning; in appealing to a phenomenalistic language as a starting point for her inquiries, the epistemologist already presupposes a good deal of science:

> Talk of subjective sense qualities comes mainly as a *derivative idiom.* [. . .] Impressed with the fact that we know external things only mediately through our senses, philosophers from Berkeley onward have undertaken to strip away the physicalistic conjectures and bare the sense data. Yet even as we try to recapture the data, in all their innocence of interpretation, we find ourselves depending upon sidelong glances into natural science. (WO, 1960b, §1)

Traditional epistemology builds on the idea that sense data are independent of our basic theories of the world. This is why a reduction of our beliefs to sense data would constitute a major epistemological achievement. Quine, however, became convinced that this presupposition is incorrect.[22]

As an example of the dependence relation between science and sense data, Quine discusses the idea that our elementary experiences are two-dimensional, an idea that Carnap also presupposed in the *Aufbau* when he wanted to assign the sense qualities in our two-dimensional visual field to points in the three-dimensional physical space order. According to Quine, however, the idea that our elementary visual experiences are two-dimensional is itself based on rudimentary science:

> The old epistemologists may have thought that their atomistic attitude toward sense data was grounded in introspection, but it was not. It was grounded in their knowledge of the physical world. Berkeley was bent on deriving depth from two-dimensional data for no other reason than the physical fact that the surface of the eye is two-dimensional.[23] (RR, 1973, 2)

[22] It should be noted that this is not Quine's only argument against sense data. See (WO, 1960, §§48–49) and, for an extensive list, Gibson (1982, pp. 157–59). Given the purposes of this chapter, I will here focus on Quine's argument against the epistemic priority of sense data.

[23] See also (WO, 1960, 2): "We may hold, with Berkeley, that the momentary data of vision consist of colors disposed in a spatial manifold of two dimensions; but we come to this conclusion by reasoning from the bidimensionality of the ocular surface, or by noting the illusion which can be

Not only do the empiricists' ideas about the two-dimensional basis of their construction depend on scientific knowledge, but even the very empiricism that underlies their attempts to construct science from sense data depends on their scientific picture of the world. According to Quine, the empiricists' claim that all knowledge is empirical can only itself rely on empirical knowledge:

> The champions of atomic sense data were seeking the unscientific raw materials from which natural science is made, but in so doing they were being guided, all unawares, by an old discovery that was the work of natural science itself. [. . .] It is the discovery that all our information about the external world reaches us through the impact of external forces on our sensory surfaces. [. . .] This is a scientific finding, open, as usual, to reconsideration in the light of new evidence.[24] (SSS, 1986i, 328)

As a result, the supposedly science-independent sense data, the so-called neutral basis for a purely epistemological foundation for science, are theoretical posits as much as the physical objects that the traditional epistemologist attempts to construct from them. For Quine, the only epistemological difference between the two is that our physicalistic conceptual scheme is what *actually* ties our experiences together: "The memories that link our past experiences with present ones and induce our expectations are themselves mostly memories not of sensory intake but of essentially scientific posits, namely things and events in the physical world" (FSS, 1995b, 15). We construct sense data only after we have acquired an object-based conceptual scheme. This is why painters have to be trained to reproduce their three-dimensional view of the world into a two-dimensional picture (GT, 1970c, 1).

The standard conception presupposes that traditional epistemology fails because we ought to despair of deducing science fully from sense data. The present reflections show, however, that Quine's rejection of traditional epistemology

engendered by two-dimensional artefacts such as paintings and mirrors, or, more abstractly, simply by noting that the interception of light in space must necessarily take place along a surface"; and (NNK, 1975d, 258): "the accepted basis of the construction, the two-dimensional visual field, was itself dictated by the science of the external world. [. . .] The light that informs us of the external world impinges on the two-dimensional surface of the eye, and it was Berkeley's awareness of this that set his problem."

[24] See also (OME, 1952, 225): "The crucial insight of empiricism is that any evidence for science has its end points in the senses. This insight remains valid, but it is an insight which comes after physics, physiology, and psychology, not before"; and (WO, 1960, 2): "The motivating insight, viz. that we can know external things only through impacts at our nerve endings, is itself based on our general knowledge of the ways of physical objects—illuminated desks, reflected light, activated retinas. Small wonder that the quest for sense data should be guided by the same sort of knowledge that prompts it."

beyond "Epistemology Naturalized" is guided by the second strategy distinguished above. For Quine, the epistemologists' quest for foundations was misguided *from the beginning*; there is no prior sense datum language, no transcendental science-independent perspective from which to validate science.[25]

2.5. Quine's Response to the Skeptic

Quine's rejection of first philosophy thus seems to be guided by an argument against transcendence, not by despair. Quine is not primarily worried about the epistemologists' ability to reconstruct science from sense data, but about their claim that sense data might provide a science-independent neutral foundation for science. This interpretation is confirmed by Quine's response to the skeptic, which, as I will show in this section, relies on the same type of reasoning.

Recall that, for Quine, traditional epistemology starts from a deep dissatisfaction with the problem of error, with "worries about our knowledge of the external world" (FSS, 1995b, 1). Now, if the standard conception were correct—if Quine's argument against traditional epistemology were indeed an argument from despair—his naturalism would constitute a surrender to the skeptic. For in despairing of reconstructing science from sense data, Quine would be despairing of the epistemologist's attempt to provide our beliefs with a proper foundation. In waking up from his Cartesian dream, in other words, Quine would be forced to admit that the skeptic was right all along; we simply ought to despair of providing our beliefs with the kind of justification the skeptic demands.

In reality, however, Quine does not admit that the skeptic has been right from the beginning. Instead of despairing of being able to answer the skeptic, he makes a move similar to the one discussed above: he argues that the skeptic *too* presupposes a good deal of science in her inquiries. Where the traditional epistemologist inadvertently relied on scientific knowledge in her talk about sense data, the skeptic cannot question science without presupposing science:

> Doubt prompts the theory of knowledge, yes; but knowledge, also, was what prompted the doubt. Scepticism is an offshoot of science.

[25] See also (WO, 1960, 3): "There is every reason to inquire into the sensory or stimulatory background of ordinary talk of physical things. The mistake comes only in seeking an implicit subbasement of conceptualization, or of language. Conceptualization on any considerable scale is inseparable from language, and our ordinary language of physical things is about as basic as language gets. [. . .] If we improve our understanding of ordinary talk of physical things, it will not be by reducing that talk to a more familiar idiom; there is none."

> The basis for scepticism is the awareness of illusion, the discovery that we must not always believe our eyes. Scepticism battens on mirages, on seemingly bent sticks in water, on rainbows, after-images, double images, dreams. But in what sense are these illusions? In the sense that they seem to be material objects which they in fact are not. Illusions are illusions only relative to a prior acceptance of genuine bodies with which to contrast them.[26] (NNK, 1975d, 258)

Skeptical questions are thus questions internal to science. According to Quine, it is science itself that shows that our evidence for science is meager; the skeptic needs to presuppose at least some theory in order to question it. The skeptic too is misguided when she believes that she can coherently doubt the reality of our beliefs from some self-sufficient science-independent perspective. Her terms too are only intelligible within a more inclusive theory of the world: "the term 'reality', the term 'real', is a scientific term on a par with 'table', 'chair', 'electron', 'neutrino', 'class', [. . .] all these are part of our scientific apparatus, our terminology, so that the only sense I can make of scepticism is that somehow our theory is wrong" (EBDQ, 1994c, 252).[27]

The question of how theory relates to evidence is an open question, but it is a question internal to science; it is an *immanent* challenge. We cannot step outside our conceptual scheme and question everything all at once. As a *transcendental* challenge skepticism simply makes no sense: "There is no such cosmic exile" (WO, 1960, 275), no self-sufficient vantage point from which to question science.[28]

[26] See also (RR, 1973, 2–3): "The skeptics cited familiar illusions to show the fallibility of the senses; but this concept of illusion itself rested on natural science, since the quality of illusion consisted simply in deviation from external scientific reality."

[27] This perspective on skeptical challenges Quine also developed in the ten years between "Two Dogmas" and *Word and Object*. See (SLS, 1954b, 229).

[28] In "Things and Their Place in Theories" (TTPT, 1981b, 21), Quine argues both that "[r]adical scepticism [. . .] is not of itself incoherent" and that in a naturalized epistemology "the transcendental question of the reality of the external world" evaporates. These remarks have puzzled many commentators. Susan Haack (1993a, 1993b), for example, has argued that these claims exemplify a fundamental ambiguity in Quine's position with respect to skepticism. If we keep in mind the immanent-transcendent distinction, however, Quine's remarks in "Things and Their Place in Theories" make perfect sense. When Quine argues that "the transcendental question of the reality of the external world" evaporates, we should note that he is explicitly talking about a *transcendental* question; that is, a question asked from a science-independent vantage point. And when Quine, on the same page, argues that "radical scepticism is not of itself incoherent," he again qualifies his claim by adding that these skeptical doubts "would still be *immanent*, and of a piece with the scientific endeavour" (my emphasis). Quine, in other words, allows skepticism when it is interpreted immanently, but dismisses it when it is interpreted transcendentally. See Verhaegh (2017c).

2.6. Reinterpreting "Epistemology Naturalized"

Let me sum up what we have established thus far. Quine's rejection of first philosophy, both before and after "Epistemology Naturalized," is not based on despair, but on his rejection of *transcendental vantage points*, his dismissal of the idea of a science-independent perspective. According to Quine, "There is no external vantage point, no first philosophy" (NK, 1969c, 127). Both the skeptic and the traditional epistemologist presuppose an Archimedean point in their inquiries. The skeptic presupposes that she can challenge science from some science-independent perspective, while the epistemologist presupposes that she can answer this challenge by reducing our theories to some science-independent sensory language.

Let me, in conclusion, examine how we might interpret "Epistemology Naturalized" in the light of these findings; that is, examine how we might make better sense of Quine's argument in the paper. I believe that the paper can be better understood if we keep in mind the distinction between immanent and transcendental inquiry. As we have seen in section 2.5, there are two ways in which one might interpret skeptical challenges. In the transcendental reading, the sceptic is seen as questioning science from some science-independent external vantage point, while in the immanent reading skepticism is a challenge from within. Quine dismisses the transcendental challenge as incoherent but admits that skeptical scenarios are live possibilities when construed immanently.

Now, since the epistemologist's project of reducing science to sense data is supposed to provide an answer to the skeptic's challenge, it admits two interpretations as well. On the one hand, one can interpret rational reconstruction as an attempt to fulfill the Cartesian dream, to ground our knowledge in some science-independent sensory language. In this transcendental reading, rational reconstruction is a project within first philosophy. As we have seen, Quine dismisses this project as incoherent because he rejects the idea of a self-sufficient sensory language.

Yet one can also interpret "rational reconstruction" as a project *internal* to science. In this reading, the project does not presuppose an external vantage point. One can just construct a phenomenalistic language, acknowledge that this language is not self-sufficient, and examine whether we can simplify our theory of the world by reducing our scientific talk to this language. In this reading, the project need not be dismissed because it presupposes an implicit subbasement of conceptualization. Rather, it fails because we ought to despair of ever being able to successfully define the empirical content of a single theoretical statement in isolation.

In this chapter, I have limited my discussion to the transcendental interpretation and argued that Quine's argument against this type of rational reconstruction is not an argument from despair. Yet the careful reader of Quine *after* "Epistemology Naturalized" will notice that Quine has never limited himself to either one of these two interpretations. Consider, for example, the following passages:

> Various epistemologists, from Descartes to Carnap, [. . .] sought a foundation for natural science in mental entities, the flux of raw sense data. It was as if we might first fashion a self-sufficient and infallible lore of sense data, innocent of reference to physical things, and then build a theory of the external world somehow on that finished foundation. The naturalistic epistemologist dismisses this dream of a prior sense-datum language.[29] (NLOM, 1995c, 462)

> My attitude toward the project of a rational reconstruction of the world from sense data is [. . .] naturalistic. I do not regard the project as incoherent, though its motivation in some cases is confused. I see it as a project of positing a realm of entities intimately related to the stimulation of the sensory surfaces, and then [. . .] to construct a language adequate to natural science. It is an attractive idea, for it would bring scientific discourse into a much more explicit and systematic relation to its observational checkpoint. My only reservation is that I am convinced, regretfully, that it cannot be done.[30] (TTPT, 1981d, 23)

Although Quine is talking about the same project in both passages, viz. reconstructing science out of sense data, the former constitutes a transcendental and the latter constitutes an immanent reading of the project. For whereas the former talks about "seeking a foundation for science," the latter talks about "positing" sense data and about bringing "scientific discourse into a much more explicit and systematic relation to its observational checkpoint." It is for this reason that Quine uses a distinct argument in each case. In the former he rejects the idea of a self-sufficient sense datum language, and in the latter he argues that the project cannot be fulfilled.

Now let me turn to "Epistemology Naturalized." Quine's goal in the paper is to convince the reader that we should abandon "creative reconstruction" and that we should examine how the construction of scientific theories "really proceeds." In order to establish this, Quine argues that there are no advantages of rational reconstruction over "straight psychology."[31] Now, when one reads

[29] See also (SSS, 1986i, 327–28).

[30] See also (PTF, 1996a, 477).

[31] As noted in chapter 1 (n. 15), Quine originally titled his paper "Epistemology naturalized; or, the Case for Psychologism." In this 1968 version of the paper, Quine's famous passage "Why not settle for psychology? Such a surrender of the epistemological burden to psychology is a move that was disallowed in earlier times as circular reasoning" reads as follows: "Why not settle for psychology?

Quine's paper with the above distinction between immanent and transcendental reconstruction in the back of one's mind, one finds that Quine is almost exclusively concerned with dismissing the advantages of rational reconstruction in its immanent reading. Quine spends almost no time on rejecting the Cartesian quest for a foundation of knowledge. He uses only a few words to argue that, with respect to the doctrinal side of epistemology, we are no farther along today than where Hume left us (EN, 1969a, 72). The implication here is that since the doctrinal project fails, the transcendental quest for foundations can be abandoned, both on its conceptual and on its doctrinal side. It is at this point that Quine could have inserted his argument against self-sufficient sensory languages; but he did not, probably because he presupposed that the reader already accepted the hopelessness of the project. The bulk of Quine's argument is concerned with dismissing the advantages of Carnap's project (pp. 72–80), a project that he interprets immanently, as he emphasizes that Carnap already saw the "Cartesian quest for certainty [. . .] as a lost cause" (p. 74).[32]

Quine, in other words, does not interpret Carnap as a first philosopher aiming to validate our scientific theories. Rather, he believes that the advantage of Carnap's project, if it were to succeed, is that it would "elicit and clarify the sensory evidence for science" (p. 74), a project that is immanent, as it will legitimize the concepts of science only "to whatever degree the concepts of set theory, logic, and observation are themselves legitimate" (p. 76). Given this immanent interpretation of Carnap's project, it is no surprise that he uses his argument from despair to dismiss it.

My suggestion, therefore, is that one should not read "Epistemology Naturalized" as an argument against traditional epistemology in its transcendental interpretation, even though some passages invite such a reading.[33] Quine and Carnap (and many other epistemologists for that matter) had

Such a surrender of the epistemological burden to psychology is a move that was denounced by Lotze, Frege, and others in the nineteenth century under the disparaging name of psychologism, partly for fear of circular reasoning" (March 1968, item 2441).

[32] See also (SM, 1965, 44): "By Carnap's day the hopelessness of grounding natural science upon immediate experience in a firmly logical way was acknowledged"; and (FSS, 1995a, 13): "Carnap's motivation was not [the] traditional quest for certainty. Rather, his goal was just a systematic integration [. . .] of our scientific concepts of mind and nature."

[33] One of the most confusing elements of "Epistemology Naturalized" is that Quine uses the term "epistemology" to denote both the Cartesian quest for certainty and the relatively innocent attempt to examine the relation between theory and evidence. In this respect, I agree with Johnsen (2005) that "Epistemology Naturalized" fails to expose Quine's views as clearly as possible. See footnote 4.

already rejected this type of first philosophy elsewhere. Rather, Quine was concerned with the type of "creative reconstruction" that continued to be an essential element of Carnap's epistemology. Quine's aim was not to show that this type of inquiry is unacceptable; he attempted to establish only that this project, regretfully, could not be fulfilled, that "[w]e must despair of any such reduction" (p. 77).

3

Naturalizing Metaphysics

3.1. Introduction

Most metaphysicians agree that we should not rest content with our ordinary ascriptions of existence. Although in everyday life and in the sciences we may freely talk about elephants, electrons, and empty sets, as philosophers we must investigate whether these objects *really* exist.[1] Carnap, as is well known, claims that these philosophical questions of existence are devoid of cognitive content. In his seminal "Empiricism, Semantics, and Ontology" (ESO), Carnap argues that existence claims only make sense internal to a linguistic framework and that we cannot ask whether an entity is "real" in an (external) framework-independent way; "reality" itself is a concept internal to a framework and as such "cannot be meaningfully applied to the [framework] itself" (Carnap, 1950, p. 207). Instead, Carnap proposes to reinterpret metaphysical questions as practical questions about which frameworks scientists ought to adopt.

Quine believes that Carnap's strict distinction between internal and external questions cannot be maintained. According to Quine, no question is purely theoretical or purely practical; just like one's decision to adapt a hypothesis in the light of new experiential data, one's decision to adopt a certain framework will be informed by both theoretical knowledge and pragmatic criteria. The question whether or not to accept a certain entity as "real" therefore *is* a meaningful question that can be answered by ordinary scientific means. Ontological questions, in other words, are "on a par with questions of natural science" (CVO, 1951c, 211).

This little stick-figure summary of the Carnap-Quine debate suggests that Quine breathed new life into the metaphysical project that was deemed meaningless by Carnap and his fellow positivists. For where Carnap rejects philosophical

[1] See, for example, Berto and Plebani (2015, p. 8): "Mathematicians talk about prime numbers; biologists talk about cross-fertile biological species; astrophysicists deal with solar flares. But, *qua* scientists, they will not typically wonder whether there are really prime numbers, species, properties or events."

existence claims as meaningless, Quine seems to restore their intelligibility by dismantling the former's internal-external distinction. Indeed, this seems to be Quine's own perspective on his debate with Carnap:

> I think the positivists were mistaken when they despaired of existence [. . .] and accordingly tried to draw up boundaries that would exclude such sentences as meaningless. Existence statements in this philosophical vein do admit of evidence, in the sense that we can have reasons, and essentially scientific reasons, for including numbers or classes or the like in the range of values of our variables. (EQ, 1968b, 97)

When quantifying over a class of entities is indispensable for the formulation of our best scientific theories, Quine argues, we are to countenance these entities as real. As a result, Quine blurs the "boundary between scientific sense and metaphysical nonsense" (CA, 1987a, 144) and concludes that Carnap and his positivistic comrades were simply "wrong if and when they concluded that the world is not really composed of atoms or whatever" (SN, 1992, 405).

The picture that Quine revived the legitimacy of philosophical existence claims is often defended in the literature as well. In his introduction to the history of analytic philosophy, for example, Avrum Stroll argues that Quine blurs "the boundary between speculative metaphysics and science, thus giving a kind of credibility to metaphysics that Carnap would never have countenanced" (2000, p. 200). Similarly, Nicholas Joll claims that "Quine saves metaphysics from positivism" (2010, §2diii) and Stephen Yablo argues that Quine, unlike Carnap, provides us with a way to attach "believable truth values to philosophical existence claims" (1998, p. 259).[2]

Yet there is something puzzling about this picture. For it does not sit well with the fact that Quine, on many occasions, *does* seem to argue against the intelligibility of metaphysical existence claims. Quine argues, for example, that he is "[n]o champion of traditional metaphysics" (CVO, 1951c, 204) and that the question "what reality is *really* like [. . .] is self-stultifying" (SN, 1992, 405). On a few occasions, Quine even argues that the "[p]ositivists were right in branding such metaphysics as meaningless" (p. 405).[3] Even more surprising from the above-sketched perspective is that Quine often appeals to the very same

[2] See also Romanos (1983). Price (2007, p. 380) aptly summarizes the above perspective on the Carnap-Quine debate by claiming that Quine is traditionally regarded as "the savior of a more robust metaphysics" by driving "a stake through the heart of ESO, [. . .] thus [dispatching] the last incarnation of the Viennese menace." See also Eklund (2013, p. 229), who argues that the above perspective dominates contemporary debates in metaontology as well. Alspector-Kelly (2001) and Price (2007, 2009) are exceptions to the received view; they were the first to suggest that the standard conception of the Carnap-Quine debate is misguided.

[3] See also, for example, Quine (SSS, 1986k, 337): "if some scientifically undigested terms of metaphysics [. . .] were admitted into science along with all their pertinent doctrine [. . .] [i]t

argument Carnap gives in ESO, viz. the argument that the notion of "reality" cannot be legitimately applied outside the system of which it is an element. We simply "cannot significantly question the reality of the external world," Quine argues, "for to do so is simply to dissociate the [term] 'reality' [. . .] from the very applications which originally did most to invest [this term] with whatever intelligibility [it] may have for us" (SLS, 1954b, 229).

Setting aside the hypothesis that Quine's metaphysical position is incoherent, one has to conclude that his views on metaphysics are subtler than has often been presupposed. Both the received view that Quine restored the intelligibility of metaphysics by dismantling Carnap's internal-external distinction *and* the opposite position that "[f]or all practical purposes, Quine does agree with Carnap about the status of metaphysical issues" (Price, 2007, p. 391) are too one-sided if we take seriously Quine's own pronouncements on the issue.[4]

In this chapter, I show how Quine is able to both blur the boundary between "natural science" and the "philosophical effusions that Carnap denounced under the name of metaphysics" (CPT, 1984a, 127–28) *and* to argue that it "is self-stultifying to ask what reality is *really* like" (SN, 1992, 405). I show that Quine allows questions about the existence of numbers, properties, etc. as legitimate scientific questions, but dismisses transcendental metaphysical questions about whether numbers and properties are "real" in a distinctively philosophical way. What I offer, then, is a detailed reconstruction of Quine's naturalistic perspective on metaphysical existence claims. I show that although Quine's position is much closer to Carnap's than the received view suggests, it still differs from the latter's position in two crucial respects.

This chapter will be structured as follows. After introducing Carnap's position in ESO as well as Quine's arguments against the former's position (section 3.2), I argue that the standard interpretation of Quine's views is incorrect because it rests on an equivocation between two different internal-external distinctions (section 3.3). Next, I show that although Quine rejects both these distinctions, he agrees with Carnap that metaphysical existence claims ought to be dismissed (section 3.4). Finally, I argue that although Quine is as skeptical as Carnap about metaphysical existence claims, there is still a significant difference between the two because they fundamentally disagree about *why* such claims ought to be dismissed (section 3.5).

would be an abandonment of the scientists' quest for economy and of the empiricists' standard of meaningfulness."

[4] To be fair, Price does not aim to provide a complete and detailed interpretation of Quine's view on metaphysics. Rather, his main aim is to show that Quine's "position on ontological commitment [. . .] *favors, or at least leaves open,* a view much closer to that of Carnap" (p. 378, my emphasis).

3.2. Internal and External Existence Claims

Carnap's problems with philosophical existence claims are deep-rooted. Already in his pre-Vienna period, he considered metaphysical disputes to be "sterile and useless" (1963a, p. 44). Influenced by the early Wittgenstein, Carnap developed the view that metaphysical theses are without cognitive content, arguing that they are pseudosentences because they "cannot in principle be supported by an experience" (1928b, p. 328). Where metaphysicians will usually agree about whether or not a certain entity is real in an everyday empirical sense, they rely on a "nonempirical (metaphysical) concept of reality" when they are involved in a philosophical dispute (p. 340).

Although Carnap never felt any qualms about these early arguments against metaphysics,[5] he later returned to the subject to explain his position once more. Carnap wrote ESO in response to critical empiricists who objected that he referred to abstract objects without having shown that they "actually exist" and argued that these empiricists "[neglect] the fundamental distinction" between ordinary and philosophical ascriptions of existence (1963a, pp. 65–66).[6]

It is in ESO that Carnap differentiates between internal and external questions in order to capture this distinction. Carnap argues that ordinary questions of existence with respect to, for example, numbers or physical things—whether there is a car in driveway, or whether there is a prime number between 31 and 41—should be viewed as questions internal to linguistic frameworks containing the rules for our use of the concepts "number" and "physical thing" and their subordinate concepts. Philosophical questions of existence about numbers and physical things, on the other hand, are to be viewed as *external* questions asked prior to the adoption of any such framework. These latter questions are the questions that Carnap's fellow empiricists had in mind when they questioned whether Carnap is justified in using a framework that quantifies over abstract objects without having shown that they actually exist. According to Carnap, however, such external questions are meaningless. For, and here his argument is similar to the one developed in his early work, the very concept of "reality" appealed to in metaphysical questions cannot be given a meaningful interpretation:

> The concept of reality occurring in [. . .] internal questions is an empirical, scientific, non-metaphysical concept. To recognize something as a real thing or event means to succeed in incorporating it into the system of things [. . .] according to the rules of the framework. From these questions we must distinguish the external question of the reality of the

[5] See Carnap's (1950, p. 215) and (1963b, p. 870).

[6] See Alspector-Kelly (2001) for a more detailed discussion of Carnap's motives in ESO.

> thing world itself. In contrast to the former questions, this question is [. . .] framed in a wrong way. To be real in the scientific sense means to be an element of the system; hence this concept cannot be meaningfully applied to the system itself. (1950, p. 207)

Concepts, according to Carnap, only make sense insofar as the rules for their use are specified within a framework. As a result, the concept of "reality" itself will only make sense within a linguistic framework, and hence philosophers fail "in giving to the external question and to the possible answers any cognitive content" (1950, p. 209).[7]

Because external questions fail to be meaningful when interpreted as theoretical, Carnap proposes that we should view them as practical questions, as matters "of practical decision concerning the structure of our language" (1950, p. 207). Rather than asking whether or not a certain entity *x* "really" exists, we should ask whether or not it is useful to adopt one or another *x*-related framework, a question that will be guided by pragmatic criteria. According to Carnap,

> [T]he introduction of [a] framework is legitimate in any case. Whether or not this introduction is advisable for certain purposes is a practical question of language engineering, to be decided on the basis of convenience, fruitfulness, simplicity, and the like. (1963a, p. 66)

As a result, although Carnap considers external questions to be devoid of cognitive content, such questions can still be given "a meaning by reinterpreting them or, more exactly, by replacing them with the practical questions concerning the choice of certain language forms" (1963b, p. 869). Carnap's internal-external distinction, in short, becomes a distinction between the theoretical and the practical when external questions are reinterpreted as questions about whether or not to adopt a certain framework.

Quine rejects Carnap's distinction between the internal and the external. In fact, he develops two arguments against the distinction: one in which he reduces it to the analytic-synthetic divide, and one in which he argues that both internal and external statements are partly theoretical and partly practical in nature.[8]

[7] Of course, Carnap's argument with respect to the concept of "reality" applies equally to philosophical notions that serve the same function. In their metaphysical inquiries, Carnap argues, philosophers might also talk about "subsistence" or the "ontological status" of an entity. These alternative philosophical notions are also without content, as philosophers have failed to explain their use "in terms of the common scientific language" (p. 209).

[8] Quine develops these arguments in "Two Dogmas of Empiricism" (TDE, 1951a, 45–46), "On Carnap's Views on Ontology" (CVO, 1951c), and "Carnap and Logical Truth" (CLT, 1954a, 132). In the second essay, Quine also provides a third argument against Carnap's distinction by reducing it to a dichotomy between category and subclass questions. According to Quine, external questions are concerned with the existence of entities expressed by a category word (e.g. "Are there *things*?"

Quine's first argument is largely negative. He argues that "a double standard for ontological questions and scientific hypotheses" requires "an absolute distinction between the analytic and the synthetic," a distinction which he famously argues to be untenable (TDE, 1951a, 43–44). In Quine's interpretation, the semantic rules of a linguistic framework are analytic, whereas internal statements are either analytic or synthetic depending on the nature of the framework in question.[9] As a result, if Carnap is correct that "statements commonly thought of as ontological are proper matters of contention only in the form of linguistic proposals," then these philosophical existence claims can only be distinguished from internal statements by appealing to an analytic-synthetic distinction (CVO, 1951c, 210).[10]

Quine's second argument is more positive, and is based on his constructive "empiricism without the dogmas" (TDE, 1951a, §6). If, as Quine maintains, science is a unified structure whose statements face experience only in clusters such that no statement is in principle immune to revision, then all statements that are relevant to science, including Carnap's linguistic proposals, will be guided by both theoretical and practical concerns. Just like the decision to adapt a hypothesis in the light of new experiential data, one's decision to adopt a certain framework will be informed by theoretical as well as practical considerations:

> Within natural science there is a continuum of gradations, from the statements which report observations to those which reflect basic features say of quantum theory. [. . .] The view which I end up with [. . .] is that statements of ontology [. . .] form a continuation of this continuum, a continuation which is perhaps yet more remote from observation. [. . .] The differences here are in my view differences only

or "Are there *numbers*?"), whereas internal questions are concerned with the existence of subclasses of them (e.g. "Are there *rabbits*?" or "Are there *prime numbers between 10 and 20*?"). He then argues that the latter distinction is trivial, because "there is no evident standard of what to count as a category" (EQ, 1968b, 92). Several scholars have argued that Quine's third argument misses the mark because the distinction between internal and external questions of existence cannot be based on the category-subclass distinction; category as well as subclass questions can be asked in both an internal and an external vein. See, for example, Haack (1976, p. 468) and Bird (1995, p. 49). For different interpretations of Quine's third argument, see Price (2009, §§4–5), where Quine's dismissal of the category-subclass distinction is read as an argument against Carnap's idea that language can be divided into frameworks; and Ebbs (2017, ch. 2), where the category-subclass distinction is interpreted as a distinction between trivial internal questions and nontrivial internal questions.

[9] E.g. internal statements in the thing-language are synthetic, whereas internal statements in the number-language are analytic. See Carnap (1950, pp. 208–9).

[10] Carnap himself also seems to have recognized the connection between the internal-external distinction on the one hand, and the analytic-synthetic distinction on the other. See Carnap (1950, p. 215 n. 5).

in degree and not in kind. Science is a unified structure, and in principle it is the structure as a whole, and not its component statements one by one, that experience confirms or shows to be imperfect. Carnap maintains that ontological questions [. . .] are questions not of fact but of choosing a convenient conceptual scheme or framework for science; and with this I agree only if the same be conceded for every scientific hypothesis.[11] (CVO, 1951c, 211)

Quine, in sum, dissolves Carnap's internal-external distinction, first by arguing that it relies on the untenable analytic-synthetic divide, and second by arguing that a better model of theory change construes all revisions as guided by both theoretical and practical considerations. In arguing that the difference between scientific and ontological claims is only gradual, Quine seems to blur the boundary between metaphysics and science, a boundary that Carnap had propagated in order to dismiss metaphysics as meaningless.

3.3. Two Distinctions

The question whether or not Quine's arguments effectively undermine Carnap's internal-external distinction has been a matter of some controversy.[12] In what follows, however, I limit my discussion to a more specific, not entirely unrelated issue, viz. the question what type of internal-external distinction Quine *aimed* to undermine. I argue that we ought to differentiate between two types of internal-external distinctions and that Quine's arguments apply to only one of them.

As we have seen in the previous section, Carnap differentiates three types of questions of existence depending on whether we are concerned with the metaphysician's perspective or with Carnap's practical reinterpretation:

(I) *Internal questions* about the existence or reality of a certain kind of entities, asked after the adoption of a linguistic framework

(E_1) *External questions* about the existence or reality of a certain kind of entities, asked before the adoption of a linguistic framework

[11] See also Quine's unpublished lecture "Mathematical Entities" (November 26, 1950, item 2958): "we find ourselves saying, with Carnap: choose your ontology as proves convenient. But I think Carnap is wrong supposing that our choice here is different in principle from, and freer then, our choice of a physical theory in the light of sense experience."

[12] See Haack (1976, §3), Bird (1995), and Glock (2002, §5) for a critical evaluation of Quine's arguments. Yablo (1998, §§5–7) and Gallois (1998, §2), on the other hand, yield a more positive verdict.

(E_2) *External questions* about whether or not it is advisable to adopt a particular linguistic framework

Although it is generally recognized that Carnap distinguishes between these three types of questions,[13] E_1 and E_2 questions are often conflated under the general heading "external questions" in discussions about the Carnap-Quine debate. As a result, some have failed to realize that we cannot speak about "*the* internal-external distinction" in general. That is, scholars often ignore the fact that ESO contains two such distinctions, depending on what type of external question one is talking about:

(I/E_1) A distinction between meaningful internal questions and metaphysical external questions without cognitive content

(I/E_2) A distinction between internal questions of a theoretical nature and external questions of a practical nature

I/E_1 primarily distinguishes between the meaningful and the meaningless, whereas the I/E_2 distinction emphasizes the difference between questions of a theoretical and questions of a practical nature.

Now, consider the question whether Quine was attacking I/E_1 or I/E_2. If one believes that Quine, in criticizing the internal-external distinction, aimed to revive metaphysical existence claims, one clearly presupposes that Quine was attacking I/E_1.[14] For if Quine had really aimed to breathe new life into the metaphysical project that was deemed meaningless by Carnap, he would have tried to show that the distinction between internal questions and E_1 questions should not be viewed as a distinction between the meaningful and the meaningless. He would have tried to show, in other words, that Carnap's E_1 questions can be given "a clear cognitive interpretation" or can be given "a formulation [. . .] in terms of the common scientific language" (Carnap, 1950, p. 209).

Yet Quine does not seem to be concerned with anything like this at all. Rather, as we shall see, there are many reasons for thinking that he was concerned with undermining not I/E_1 but I/E_2.[15] Quine's arguments against Carnap's distinction provide the first reason. Recall that Quine first reduces the internal-external

[13] See, for example, Eklund (2013, p. 237): "Carnap is actually drawing a *tripartite* distinction: between questions internal to a framework, questions about which framework we should choose to employ, and the pseudo-questions."

[14] See, for example Haack (1976, §3.1) and Bird (1995, §2), where Quine's arguments are evaluated in terms of their effectiveness in undermining I/E_1.

[15] Eklund (2013) correctly suggests that Quine attacks I/E_2, although he does not argue for this claim.

distinction to the analytic-synthetic divide and then argues that both scientific hypotheses and Carnap's linguistic proposals are guided by theoretical as well as practical considerations. Now, if Quine were really aiming to undermine I/E_1, then neither of these arguments makes sense. Since Carnap rejects E_1 questions as meaningless, they are neither analytic nor synthetic; an argument against the analytic-synthetic distinction therefore has no relevance if one aims to undermine I/E_1. A similar conclusion can be drawn about Quine's second argument. Quine's claim that the distinction between the theoretical and the practical is a matter of degree, not kind, is not relevant had he targeted I/E_1, since Carnap does not view E_1 questions as practical questions.[16]

If, on the other hand, we interpret Quine as arguing against Carnap's I/E_2 distinction, his arguments make perfect sense. For the lack of a sharp distinction between the analytic and the synthetic seriously undermines Carnap's attempt to draw a distinction between "the acceptance of a language structure and the acceptance of an assertion formulated in the language" (1950, 215). Similarly, if both scientific hypotheses and linguistic proposals are guided by theoretical as well as practical considerations, Carnap cannot uphold his claim that the two can be distinguished because internal questions are theoretical and E_2 questions are practical. Quine, in other words, both shows that an I/E_2 distinction cannot be maintained and develops a positive theory in which the distinction between the theoretical and the practical is a matter of degree.

A second reason for thinking that Quine was concerned with undermining I/E_2 instead of I/E_1 is the way in which he describes Carnap's external questions. In his critical papers on Carnap's distinction, Quine consistently refers to those questions as "linguistic proposals." In "Two Dogmas," for instance, he argues that Carnap sees ontological questions as concerned with "choosing a convenient language form, a convenient conceptual scheme or framework for science" (TDE, 1951a, 45).[17] Describing external questions in this way only makes sense if Quine has in mind Carnap's E_2 questions. E_1 questions, after all, are certainly not linguistic proposals; they are metaphysical questions concerning the reality of a certain class of entities.

[16] See also Price (2009, p. 326): "Quine's claim is that there are no purely internal issues, in Carnap's sense. No issue is ever entirely insulated from pragmatic concerns about the possible effects of revisions of the framework itself. [. . .] Quine's move certainly does not restore the non-pragmatic external perspective required by metaphysics. In effect, the traditional metaphysician wants to be able to say, 'I agree it is useful to say this, but is it true?' Carnap rules out this question, and Quine does not rule it back in."

[17] See also Quine's (ML1, 1950a, 208), (CVO, 1951c, 210) and (CLT, 1954a, 132), where Carnapian external questions are described as a "linguistic convention devoid of ontological commitment," as a "linguistic proposal" and as a "matter [. . .] of linguistic decision."

Finally, my interpretation of Quine's aims in rejecting the internal-external distinction is supported by the background of the debate between Carnap and Quine. I have already noted that Carnap wrote ESO in order to respond to critical fellow empiricists who objected that he referred to abstract objects without having shown that they "actually exist." Now, as it turns it out, Quine was one of those critics.[18] In the late 1930s, Quine developed his criterion of ontological commitment, according to which we are committed to an entity "if and only if we regard the range of our variables as including such an entity" (LAOP, 1939a, 199).[19] Carnap, who claimed to accept Quine's criterion,[20] however still maintained that his talk about abstract entities should be seen as "a practical decision like the choice of an instrument" (1947, §10). From Quine's perspective, therefore, Carnap was dodging his ontological commitments. That is, although Carnap accepted Quine's "standard for judging whether a given theory accepts given alleged entities" (CVO, 1951c, 205), he still did not acknowledge that he was committed to abstract objects because he viewed his "acceptance of such objects [as] a linguistic convention distinct somehow from serious views about reality (WO, 1960b, 275).[21]

If we take this background into consideration, it becomes clear that when Quine attacked Carnap's internal-external distinction in the early 1950s, he was not concerned with the latter's claim that the traditional metaphysician's questions are devoid of cognitive content, i.e. with the claim that E_1 questions are meaningless. Rather, his job was to argue that there is no proper distinction between the ontological commitments internal to a framework and the linguistic conventions upon which our framework choices are based, i.e. the distinction between internal questions on the one hand and E_2 questions on the other. Indeed, when Quine first learned about the internal-external distinction in a 1949 letter from Carnap, he scribbled on the back of this letter: "When are

[18] In his "Intellectual Autobiography," Carnap lists Quine as one of the philosophers who rejected his way of speaking as "a 'hypostatization' of entities" (1963a, p. 65).

[19] For a reconstruction of the genesis of Quine's criterion, see Decock (2004). I also briefly sketch the development of the criterion in section 5.2.

[20] See Carnap (1950, p. 214 n. 3).

[21] Quine's discontent with Carnap's position can be traced back at least to 1937, when he, in a lecture on nominalism, suggests that although Carnap succeeds in avoiding metaphysical questions by rejecting them as meaningless, he does not "provide for reduction of all statements to statements ultimately about tangible things, matters of fact," and thereby fails to show how we can keep "our feet on the ground—avoiding empty theorizing" (October 25, 1937, item 2969, my transcription). See Mancosu (2008, pp. 28–29). See also Alspector-Kelly (2001, §3): "As Quine understands it, Carnap endorsed Quine's criterion of ontological commitment. [. . .] Nonetheless, Carnap did not take himself to be committed to abstract entities, and so did not take himself to be a Platonist, despite the fact that he quantified over abstract objects. Nor did he have any plan to show that such quantification can be avoided."

rules really adopted? Ever? Then what application of your theory to what I am concerned with (language now)? [. . .] Say frameworkhood is a matter of degree, & reconciliation ensues" (QCC, ca. August 1949, 417). Whether or not this is consistent with Carnap's intentions, therefore, Quine from the very beginning interpreted Carnap's distinction as one between questions internal to a framework and questions regarding the choice of the framework itself.

Quine, in sum, was not out to attack the I/E_1 distinction; he was concerned with undermining Carnap's I/E_2 distinction. Quine did not aim to restore the legitimacy of metaphysics, but rather to criticize the Carnapian view that "statements commonly thought of as ontological are proper matters of contention only in the form of linguistic proposals" (CVO, 1951c, 210).[22]

3.4. Quine on Metaphysical Existence Claims

If I am right in claiming that Quine aims to undermine I/E_2 instead of I/E_1 in his critical papers on Carnap's internal-external distinction, then a question that remains to be answered is what Quine's position on I/E_1 *is*. After all, the claim that Quine aimed to criticize Carnap's I/E_2 distinction does not imply anything about Quine's views about the tenability of the I/E_1 dichotomy. In the remainder of this chapter, I address this latter question.

Let me start with Quine's views on E_1 questions. From the very beginning of his philosophical career, Quine was skeptical about metaphysical existence claims. In one of his early (1938) letters to Carnap, for example, he already characterized "metaphysical expressions" as "devoid of denotation, truth, and falsehood" (QCC, February 15, 1938, 247–48). This attitude did not change when he developed his criterion of ontological commitment. For Quine always made clear that his criterion is only concerned with questions of existence from the point of view of a given language, or as he phrases it in "On What There Is":

> We look to bound variables in connection with ontology not in order to know what there is, but in order to know what a given remark or doctrine, ours or someone else's, *says* there is. (OWTI, 1948, 15)

The traditional metaphysician's question of existence, in other words, falls outside the scope of his theory of ontological commitment. Questions about what

[22] Given this misunderstanding, it is not surprising that some scholars have concluded that "Quine's criticisms leave Carnap's central points untouched" (Bird, 1995, p. 41). For even if Carnap's central point was to distinguish between internal questions and E_1 questions—a claim that might be doubted given the background of the Carnap-Quine debate—Quine was simply not concerned with criticizing that distinction.

a theory *says* there is, after all, are internal questions and not E_1 questions.[23] But whereas Carnap always advertised his dismissal of E_1 questions, Quine, in those early stages of his career, limited himself to brief remarks in his letters to Carnap and personal notes.[24] Quine's first *published* remarks concerning his views about E_1 questions, as we shall see shortly, are from the 1950s.

Quine's early reservations about rejecting metaphysical questions of existence are explained by the fact that he took himself to be explicating the elements of traditional metaphysics that *are* legitimate. Quine believed that his use of the concept "ontology" in his theory of ontological commitment had been "nuclear to its usage all along" (1951c, 204). So although Quine, like Carnap, proposed to *reinterpret* the traditional metaphysician's questions, he did not, like Carnap, explicitly distance himself from the concepts used by those traditional metaphysicians: "meaningless words," he claimed, "are precisely the words which I feel freest to specify meanings for" (CVO, 1951c, 203).

This difference is illustrated by Carnap's and Quine's approaches to the question of nominalism. Where Carnap rejected the issue of nominalism "as meaningless because metaphysical," Quine believed that "the problem of universals [. . .] *can* be given [. . .] an important meaning" (NO, 1946a, 9, my emphasis):

> As a thesis in the philosophy of science, nominalism can be formulated thus: it is possible to set up a nominalistic language in which all of natural science can be expressed. The nominalist, so interpreted, claims that a language adequate to all scientific purposes can be framed in such a way that its variables admit only concrete objects, individuals, as values. (DE, 1939b, 708)

Carnap agreed that Quine's reinterpretation of nominalism "is a meaningful problem" but doubted whether it is "advisable to transfer to this new problem in logic or semantics the label 'nominalism,'" because the concept stems from the "old metaphysical problem" (1947, p. 43). Again, Carnap and Quine agreed about which types of questions are legitimate, but disagreed about whether or not those questions ought to phrased using the traditional metaphysician's concepts.[25] Carnap believed it to be safer to introduce new concepts, whereas Quine wanted to emphasize that he was explicating those elements of the traditional metaphysician's question that *are* significant.[26]

[23] To be more precise, questions about what a theory says there is are partly internal and partly E_2, according to Quine, because he believes no statement to be purely theoretical or purely practical.

[24] See, for example, Quine's letter to Carnap from May 1, 1947: "most metaphysical statements simply mean nothing to me" (QCC, May 1, 1947, 410).

[25] Cf. Alspector-Kelly (2001, §7).

[26] See also, for example, Quine's letter to Carnap from January 5, 1943, in which Quine talks about there being a "kernel of technical meaning in the old controversy about reality or irreality of

So although their positions seem to differ greatly at surface level, Carnap's and Quine's views on E_1 questions were actually remarkably similar in the 1930s and 1940s. Both dismissed metaphysical existence claims but accepted Quine's theory of ontological commitment as well as his reinterpretation of nominalism. They differed only on whether Quine's theories were to be viewed as faithful explications of the traditional metaphysician's questions. Carnap, as we have seen, proposed to "replace" them with "practical questions concerning the choice of certain language forms" (1963a, p. 869), whereas Quine believed that there is no I/E_2 distinction and hence proposed to replace them by questions about "the ontological commitments of a given doctrine or body of theory" (CVO, 1951c, 203). That is, Carnap and Quine agreed that E_1 questions ought to be rejected but differed about whether to *reinterpret* those questions as either I/E_2 questions (like Quine) or as purely E_2 questions (like Carnap). Quine's reticence to be explicit about his views on E_1 questions has probably contributed significantly to the misunderstanding that he was aiming to restore the intelligibility of metaphysical existence claims.[27]

In later work, however, Quine *did* become more explicit about the difference between his views on ontology and the questions asked by traditional metaphysicians. He came to accept that Carnap was right in claiming that philosophers who treat questions of existence "as a serious philosophical problem [. . .] do not have in mind the internal question" (Carnap, 1950, p. 209).[28] That

universals" (QCC, January 5, 1943, 295). Somewhat related, in 1947 Goodman and Quine published their joint paper "Steps towards a Constructive Nominalism" (SCN, 1947). When Goodman, in a letter, proposed to dub the joint position he and Quine defended "particularism" (May 18, 1947, item 420), Quine responded that they should stick with "nominalism" because it is "a shame to disavow a noble tradition when we are squarely in line with it" (June 12, 1947, item 420). See also Mancosu (2008, p. 42).

[27] In chapter 5, I reconstruct the development of Quine's early views about metaphysics in more detail.

[28] In "Ontology and Ideology," for example, Quine distinguishes between "ontology" and "absolute ontology" and defines the latter as the study of "what there really is" (OI, 1951b, 14). "Ontology and Ideology" is a response to Gustav Bergmann's "A Note on Ontology" (1950). This paper is generally ignored in discussions about Quine's views on ontology, but Quine himself viewed the paper as some sort of counterweight to his response to Carnap. In his autobiography, Quine remarks:

> I went to a meeting of Carnap's seminar in Chicago [. . .] to join issues with him on [. . .] ontology. [. . .] Feigl was there and requested my piece for his journal, *Philosophical Studies*. It appeared as "On Carnap's Views on Ontology." It had been preceded by my "Ontology and Ideology," a requested response to Gustav Bergmann's contention that the predicates of a theory commit us directly to an ontology of corresponding properties. Between Bergmann's excessive acknowledgments of abstract objects and Carnap's disavowals of them, I was midway. (TML, 1985a, 226–27)

is, he explicitly recognized that there are two ways to understand questions of existence, an ordinary one (e.g. "Are there numbers?") and a philosophical one (e.g. "Are there *really* numbers?"). The traditional metaphysician, Quine argued, is not interested in questions of ontological commitment but rather wants to "inquire into the *absolute* correctness of a conceptual scheme," wants to answer "[t]he fundamental-seeming philosophical question, How much of our science is merely contributed by language and how much is a genuine reflection of reality?" (IOH, 1950b, 78–79), wants to know "what reality is *really* like" (SN, 1992, 405), or, in Kantian terms, "whether or in how far our science measures up to the *Ding an sich*" (TTPT, 1981b, 22). In response to these questions, Quine now explicitly argued against traditional metaphysics. According to Quine, any inquiry into the absolute correctness of a conceptual scheme is "meaningless" (IOH, 1950b, 79) and any question about what reality is really like is "self-stultifying" (SN, 1992, 405).

Yet not only did Quine become more explicit about his position with respect to E_1 questions, he also started to develop an argument against them. And just as his views on E_1 questions are similar to Carnap's, his argument against those questions is in Carnapian spirit as well. As we have seen, Carnap's argument against E_1 questions relies on the idea that the very concept of "reality," which plays an important role in the metaphysician's question about whether a certain object *really* exists, cannot be meaningfully applied outside the framework of which it is an element. If we now replace Carnap's talk about frameworks with Quine's holistic picture of science as a "man-made fabric which impinges on experience only along the edges" (TDE, 1951a, 42), we get a very similar argument. In the early 1950s, Quine starts to argue that key philosophical concepts like "reality" cannot be divorced from their everyday scientific applications. When the traditional metaphysician asks us what reality is *really* like, Quine argues, she "dissociate[s] the [term] 'reality' [. . .] from the very applications which originally did most to invest those terms with whatever intelligibility they may have for us" (SLS, 1954b, 229). According to Quine, "[t]here is no deeper sense of 'reality' than the sense in which it is the business of science itself, [. . .] to seek the essence of reality" (QWVO, 1996b, 348). When the traditional metaphysician asks us about the true nature of reality, in other words, she presupposes that we can separate the term "reality" from its ordinary scientific use. According to Quine, however, this cannot be done.[29]

[29] The argument first occurs in Quine's "On Mental Entities," where he speaks about "the ordinary usage of the word 'real'" (OME, 1952, 225).

3.5. Scientific Sense and Metaphysical Nonsense

Let me sum up what we have established thus far. I have argued that we ought to distinguish between two types of internal-external distinctions: I/E_1 and I/E_2. Quine should not be viewed as aiming to attack I/E_1, thereby breathing new life into the metaphysical project that was deemed meaningless by Carnap, because he was solely concerned with undermining I/E_2. In fact, Quine's perspective on E_1 questions, even in the early stages of his career, is remarkably similar to Carnap's. For although Carnap and Quine disagree about how to *reinterpret* E_1 questions, they both reject these questions and use similar arguments to show why they ought to be dismissed.

In the light of these facts, one might conclude that Quine is committed to an I/E_1 distinction himself. That is, one might conclude that Quine too is committed to a distinction between scientific sense and metaphysical nonsense.[30] This would be a too hasty conclusion, however, as Quine always maintained that he was "blurring [. . .] the boundary" between "natural science" and the "philosophical effusions that Carnap denounced under the name of metaphysics" as well (CPT, 1984a, 127–28).[31]

How can this be? How can Quine both dismiss metaphysical existence claims and at the same time reject Carnap's distinction between science and traditional metaphysics? I believe that the answer can be found in Quine's rejection of the criterion of significance that underlies Carnap's distinction. For although Quine sometimes uses the term "meaningless" in dismissing metaphysics,[32] there is ample evidence that he does not, like Carnap, subscribe to a strict, philosophically potent distinction between the meaningful and the meaningless. In *From Stimulus to Science*, for example, Quine explicitly dismisses the positivistic thesis that a sentence "is meaningless unless it has *empirical content*" (FSS, 1995a, 48). Because of his holistic view that no single statement has "distinctive content of its own" (CPT, 1984a, 125–26), Quine argues no strict criterion of empirical content can be drawn. Against the suggestion that a sentence *q* has empirical content if and only if it is a supporting member of a set of sentences *Q* with critical mass such that if *q* is withdrawn it deprives *Q* of empirical content, for

[30] Indeed, I used to think this myself. See Verhaegh (2015, ch. 3), in which I argue that Quine himself presupposes something like an I/E_1 distinction in his disquotational theories of truth and reference. I am greatly indebted to Peter Hylton and Gary Ebbs for pressing some of the issues discussed in the present section, which has led me to conclude differently.

[31] See also Quine (TDE, 1951a, 20), where he speaks about "a blurring of the supposed boundary between speculative metaphysics and natural science" and (CA, 1987a, 144), where he dismisses the "boundary between scientific sense and metaphysical nonsense" as "dubious."

[32] For instance, when he claims that the "[p]ositivists were right in branding [. . .] metaphysics as meaningless" (SN, 1992, 405).

example, he argues that on this criterion any sentence whatsoever will have empirical content trivially:

> Any sentence, even Russell's 'Quadruplicity drinks procrastination', is a supporting member of a set that implies an observation categorical. Let us abbreviate Russell's sentence as '*q*', and some observational categorical as '*c*'. The two-member set {'*q*', '*q* ⊃ *c*'} implies '*c*', but the one-member set {'*q* ⊃ *c*'} does not. So Russell's sentence is a supporting member of {'*q*', '*q* ⊃ *c*'}. (FSS, 1995a, 48)

Quine, in other words, knows no way to strictly distinguish between statements with and statements without empirical content. More important given our purposes here, however, is that Quine insists that even if it *were* possible to draw such a strict distinction, he would not want to employ it as a distinction between the meaningful and the meaningless. For even statements that are completely without empirical content, Quine argues, should not be dismissed as meaningless:

> Even if I had a satisfactory notion of shared content, I would not want to impose it in a positivist spirit as a condition of meaningfulness. Much that is accepted as true or plausible in the hard sciences, I expect, is accepted without thought of its joining forces with other plausible hypotheses to form a testable set. [. . .] Positivistic insistence on empirical content could, if heeded, impede the progress of science. (FSS, 1995a, 48–49)

Quine, in sum, cannot dismiss metaphysical existence claims as meaningless because he does not subscribe to a philosophical criterion of significance at all.[33]

On what grounds, then, *does* Quine dismiss metaphysical existence claims? The answer is surprisingly simple. As early as 1937, Quine provides a solution to this question in one of his notebooks. Metaphysical existence claims are not meaningless, they are merely *useless* from a scientific perspective:

> Meaninglessness must be abandoned as meaningless—at least insofar as it might be used against metaphysics. Even supposing we would make sense ultimately of an operational criterion, this would rule out all the non-intuitionistic part of math[ematics] also. But we keep [the] latter, because *useful* algorithmically for science. We discard metaphysics because useless for science. If part of met[aphysics] became

[33] See also Quine's 1993 letter to Putnam: "I don't censor à la Carnap nor demarcate à la Popper" (February 1, 1993, item 885). For a more extensive argument for the thesis that "there is simply no idea of nonsense that is defensible in Quinean terms which will play an significant philosophical role," see Hylton (2014b).

> useful for science, we might use it on same grounds as non-constructive math[ematics]. (April 2, 1937, item 3169, my transcription)

The idea that metaphysical existence claims are not meaningless but useless for science is something Quine held on to until the end of his career. In his unpublished Notre Dame Lectures, for instance, Quine argues that although metaphysical questions are "technically meaningful," they are "pragmatically empty and need not arise" (May 11–15, 1970, item 2971, my transcription). Similarly, in his 1986 paper "The Way the World Is," Quine proposes that portions of science without empirical content (e.g. higher set theory) should be accepted as meaningful because they "deal with questions that can be formulated in the same vocabulary and the same idioms that are also *useful*, in other combinations, elsewhere in science." He is not willing, however, to grant metaphysics a similar position: "There are further reaches of discourse [. . .] for which not even these claims to a scientific status can be made. One thinks here of bad metaphysics" (WW, 1986h, 169, my emphasis). Presumably the reason is precisely that metaphysics *cannot* "be formulated in the same vocabulary" as science. The traditional metaphysician's terms are useless because in asking us about the true nature of reality, she uses the term "reality" in a nonstandard way. For Quine, asking what reality is *really* like independently of our scientific system of the world is "like asking how long the Nile really is, apart from parochial notions of miles or meters" (SN, 1992, 405). Just as the term "length" is useless if it is divorced from related notions like "mile" and "meter" and some standards of measurement, the term "reality" is useless when one purports to use it in a way that is divorced from our theory of the world and our scientific standards.

Quine, in conclusion, can both dismiss metaphysical existence claims and, at the same time, reject Carnap's strict distinction between science and metaphysics because he, unlike Carnap, refuses to appeal to a strict criterion of significance. Where Carnap requires a sharp "metalinguistic tool" to distinguish between "the meaningful claims of science and the meaningless claims of metaphysics" (Ebbs, 2011a, p. 198), Quine argues that we can get rid of metaphysical existence claims *without* dismissing them as meaningless. Quine, in other words, does not require a metalinguistic criterion to get rid of metaphysical existence claims; for from "within our total evolving doctrine" (WO, 1960b, 25), we simply have no *use* for metaphysical hypotheses.[34]

[34] Indeed, in a set of notes for a conference in Oberwolfach, Quine explains his position by referring to Pierre-Simon Laplace's famous reply to Napoleon about the existence of God: "What of mngless. of metaphysics? Well, just danglers, not contributing to joint implications of observ'n. 'Je n'ai pas besoin de cette hypothèse' " (March 1987, item 2928, my transcription).

3.6. Conclusion

Carnap's "Empiricism, Semantics, and Ontology" contains two internal-external distinctions: one between scientific sense and metaphysical nonsense and one between questions internal to a framework and questions regarding the choice of the framework itself. I have argued that although Quine dismisses both distinctions—the former because there is no strict philosophically potent criterion of significance, the latter because there is no strict distinction between the theoretical and the practical—we should not interpret Quine as reviving the metaphysics that was deemed meaningless by Carnap. For although Quine does not reject metaphysics *as meaningless,* he dismisses metaphysical existence claims because they are useless from a scientific point of view. In one sense, therefore, Quine and Carnap "are playing for the same team" (Price, 2009, p. 323 n. 1). In another sense, however, they are not. For in blurring the "supposed boundary between speculative metaphysics and natural science" (TDE, 1951a, 20), Quine has effectively shown that we can get rid of metaphysics without relying on a metalinguistic criterion to distinguish between what can and what cannot be meaningfully said. Where Carnap was always actively concerned with developing a definitive weapon against traditional metaphysics, refusing even to use the traditional metaphysicians' concepts when reinterpreted, Quine was not especially worried about metaphysics. Quine dealt with metaphysical existence claims as he dealt with all useless statements: he simply dismissed them from within.

4

In Mediis Rebus

4.1. Introduction

Quine's arguments against traditional epistemology (chapter 2) and traditional metaphysics (chapter 3) show that his naturalism consists of two components: the principled rejection of *transcendental* perspectives, and the adoption of a perspective *immanent* to our scientific conceptual scheme:

(NT) *No Transcendence*: the rejection of any detached science-independent perspective on reality.

(SI) *Scientific Immanence*: the prima facie acceptance of our inherited scientific theories and methods.[1]

[1] Quine himself also uses the terms "immanence" and "transcendence" to characterize his naturalism. See, for example, (RA, 1994d, 230): "the immanent is that which makes sense within naturalism, *in mediis rebus*, and the transcendent is not"; (TTPT, 1981b, 22): "Transcendental argument, or what purports to be first philosophy, tends generally to take on rather this status of immanent epistemology insofar as I succeed in making sense of it"; and (SN, 1992, 406): "My global structuralism should not [. . .] be seen as a structuralist ontology. To see it thus would be to rise above naturalism and revert to the sin of transcendental metaphysics."

It should be noted that Quine also uses the terms "immanence" and "transcendence" for more two more "mundane" (WDWD, 1999a, 164 n. 7) purposes. In *Philosophy of Logic*, Quine introduces an immanence-transcendence distinction for grammatical categories. A category is immanent when it is defined for a particular language, e.g. the class of *der*-words in German grammar, and transcendent when it is defined for languages generally (PL, 1970a, 19–22). Another "mundane" use of the immanence-transcendence dichotomy can be found in (IV, 1991a, 242), where Quine distinguishes between metatheories which are and metatheories which are not expressed in the language of the object theory. In describing his ideas about the relation between science and philosophy, however, Quine is not referring to these distinctions, as they are "irrelevant to issues of naturalism" (WDWD, 1999a, 164 n. 7). In what follows, therefore, I will limit my examination to what Quine calls his "august" (as opposed to his "mundane") immanence-transcendence distinction (RA, 1994d, 230).

Although we have primarily focused on NT in the preceding chapters, NT and SI together deliver some of Quine's most characteristic theses. Epistemologically, NT entails that we ought to abandon "the Cartesian dream of a foundation for scientific certainty firmer than scientific method itself" (PT, 1990a, 19) and SI implies that our scientific theories do not require "any justification beyond observation and the hypothetico-deductive method" (FME, 1975a, 21). Metaphysically, NT shows that we ought to dismiss the transcendental question of "what reality is *really* like" (SN, 1992, 405), whereas SI implies that ontological questions are "on a par with questions of natural science" (CVO, 1951c, 211).

Although NT and SI appear to be two sides of the same coin, they are logically independent theses. A skeptic, for example, could accept NT and deny SI; she might insist that we are not justified in accepting our theories about the world from either a philosophical or a scientific perspective.[2] Conversely, many present-day nonnaturalists will presumably accept some version of SI, granting that philosophers should at least start out presupposing that our scientific theories and methods are largely correct, yet deny that there is no distinct philosophical perspective from which those theories and methods might be evaluated.

Still, I submit that NT and SI contain a common core. In this chapter, I argue that both theses are based on Quine's view that "[t]he naturalistic philosopher begins his reasoning within the inherited world theory as a going concern" (FME, 1975a, 72). I examine the nature of this commitment and show that it has far-reaching philosophical consequences—consequences which are often ignored in contemporary discussions and, accordingly, have led commentators to misinterpret Quine's position. In this chapter, in other words, I provide a positive characterization of Quine's core commitment: I show that to be a naturalist is to work from within.

This chapter is structured as follows. I start with an analysis of Quine's own defense of naturalism in "Five Milestones of Empiricism" and argue that he believes that his position is supported by three commitments: empiricism, holism, and realism (section 4.2). Next, I show that Quine's defense raises some questions and I argue that these issues can be resolved if we take seriously his view that "we must work within a growing system to which we are born" (section 4.3). Although this seems to be a relatively innocent thesis about the nature of inquiry, I argue that the view has far-reaching philosophical consequences—consequences which shed new light on the nature of SI (sections 4.4–4.5). I end this chapter with a discussion of NT, showing how Quine's commitment to

[2] Quine, of course, would argue that one cannot coherently accept NT while denying SI. See section 2.5.

"working from within" illuminates his dismissal of transcendental perspectives as well (sections 4.6 and 4.7).

4.2. Three Commitments

NT and SI provide us with an abstract characterization of Quine's naturalism. Yet the theses should not be viewed as philosophical dogmas unsupported by any further arguments; if they were, Quine would be vulnerable to the objection that his naturalism itself is a transcendental extrascientific thesis.[3]

In "Five Milestones of Empiricism" (1975a), Quine argues that naturalism is a stage in the development of empiricist philosophy. At several points in the past two centuries, Quine argues, empiricism has taken a turn for the better; and the (for now) final milestone of empiricism is naturalism. In the paper, Quine argues that his naturalism is supported by three commitments: empiricism, holism, and realism. Let me discuss the three commitments in turn.

Empiricism: The first commitment supporting Quine's naturalism as defined by NT and SI is pretty straightforward. If anything, Quine's position is bolstered by the empiricist thesis that all our information about the world ultimately comes from sense experience. After all, "Five Milestones" presents naturalism as a distinct milestone in the development of empiricist philosophy.

Empiricism bears on Quine's naturalism in two distinct ways. First, it supports NT because it rules out many purported extrascientific sources of knowledge. Traditionally, many philosophers have aimed to base our scientific theories on an indubitable a priori foundation. According to Quine, however, the empiricist can dismiss these attempts as illegitimate; there is no reason to believe that the rationalist's a priori "truths" are actually true.[4] Empiricism thus supports NT because it simply dismisses any distinctively philosophical question about the a priori foundations of science. Second, empiricism supports SI in providing us with an explanation of *why* we should accept our inherited scientific theories and

[3] For the objection that naturalism is self-refuting because it is itself not supported by our best scientific theories, see Almeder (1998, p. 64), Moser and Yandell (2000, p. 10), and Macarthur (2008, p. 10).

[4] See, for example, Quine's "Lectures on David Hume's Philosophy" (LDHP, 1946c, 54–59). After reconstructing Descartes's account of self-evidence in mathematics and philosophy, Quine asks: "Why should the self-evidence of mathematical axioms be a guarantee of their *truth*, rather than merely a compulsion to belief—possibly mistaken belief—on our part? And similarly for any other self-evident truth." According to Quine, especially the set-theoretical paradoxes show that seemingly self-evident axioms can turn out to be false. See (WRML, 1941b).

methods. If one agrees with Quine that science is our best attempt to systematically account for our sensory input (more on Quine's notion of "science" in section 4.5), then a commitment to empiricism implies that one should at least start out one's inquiries presupposing that our scientific theories and methods are largely correct.

Empiricism, however, is not just a philosophical dogma; it is itself supported by our best scientific theories: "it is a finding of natural science itself, however fallible, that our information about the world comes only through impacts on our sensory receptors" (PT, 1990a, 19). Empiricism, for Quine, is simply our best scientific theory about our sources of knowledge, as is exemplified by the fact that he believes it to be at least possible that scientists would one day find out that there are other sources of knowledge as well.[5] Of course, the justificatory structure here is somewhat circular: the respect for science that is embodied in Quine's naturalism is supported by empiricism, whereas empiricism itself is finding of science. It is characteristic of Quine's naturalism, however, that he has no problems with such circularity; since there is no extrascientific perspective, we cannot but presuppose science in justifying our prima facie acceptance of science.[6]

Quine's empiricism is relatively strict. For Quine, concepts concerning mind and language are empirically acceptable only when we are able to provide them with behavioristic definitions. The intuitive clarity of notions like "meaning" and "synonymy," for example, does not suffice to allow their use in our best scientific theories of the world; if we cannot explicate these concepts unambiguously in terms of behavioral dispositions, we should simply do without them. Indeed, behaviorism is so important to Quine that he considers adding it to his list of empiricist milestones. He refrains from doing so, however, because he sees behaviorism "as integral to naturalism" (PPE, 1975b, 37).[7]

[5] See (SSS, 1986i, 328): "There is no telepathy, clairvoyance, revelation, or extrasensory perception. This is a scientific finding, open, as usual, to reconsideration in the light of new evidence."

[6] According to Hylton, this way of reasoning is even the most characteristic feature of Quine's naturalism: "How do we know that the methods and techniques of natural science are our best source of knowledge about the world? Quine's predecessors within the analytic tradition [. . .] might at this point start [. . .] invoking philosophical ideas. [. . .] Quine, by contrast, insists that the naturalistic claim [. . .] too must be based on natural science. (If this is circular, he simply accepts the circularity.) This is the revolutionary step—naturalism self-applied, as it were" (2014a, p. 150).

[7] Although there are certainly connections between Quine's behaviorism and the behaviorism that dominated mid-twentieth-century psychology, the two should not be conflated. See Decock (2010), Føllesdal (2011), Becker (2013, pp. 232–35), Verhaegh (forthcoming-c), and, for a historical reconstruction, Verhaegh (forthcoming-d). See also Hylton (2007, §4.4), who argues that Quine's behaviorism is nothing more than the thesis that any theory of language should start from the "undeniable" fact that "language is learnt by infants who receive information about the world only

Holism: In "Five Milestones of Empiricism," Quine not only presents his naturalism as a distinct stage in the history of empiricism, but also gives us a glimpse of what he believes to be the main commitments supporting his position. After characterizing naturalism as the (for now) final milestone of empiricism, he distinguishes "two sources" of naturalism, the first of which is evidential holism:

> Naturalism has two sources, both negative. One of them is despair of being able to define theoretical terms generally in terms of phenomena, even by contextual definition. A holistic or system-centered attitude should suffice to induce this despair. (FME, 1975a, 72)

Evidential holism, as we have seen in sections 2.2 and 2.3, is the thesis that typical theoretical sentences have no distinctive empirical content of their own; only clusters of theory are inclusive enough to imply observable consequences. Whenever we are confronted with an observation contradicting our best scientific theories, "we are free to choose what statements to revise and what ones to hold fast" in restoring consistency between theory and evidence (EESW, 1975c, 230). More precisely, evidential holism is a thesis about the logical relation between theory and evidence; or, in Quinean terms, about the relation between clusters of theoretical sentences and observation categoricals.[8] The logical relation between theoretical sentences and observation categoricals can be best described by what might be called a prediction thesis and a falsification thesis:[9]

through their sensory stimulations." Hylton also credits Katz, "one of Quine's most vehement critics," for having appreciated this. See Katz (1990, pp. 179–80):

> I shall [. . .] make no objection to Quine's statement that "the behaviorist approach is mandatory." The behaviorism he has in mind here is not the dreaded reductive doctrine of days gone by, but merely a way of putting the study of language on a par with other sciences by requiring the linguist's theoretical constructions to be justified on the basis of objective evidence in the form of overt behavior of speakers. [. . .] Quine's behaviorism is thus a behaviorism one can live with.

The question, according to Katz, is thus not whether Quine's approach is correct, but whether his claims about *what* can be gleaned from overt behavior are correct.

[8] For a definition of observation categoricals, see chapter 2, footnote 16. It should be noted that Quine has not always explicated observational predictions in terms of observation categoricals. See, for example, (WO, 1960) and (EESW, 1975c).

[9] The terms "prediction thesis" and "falsification thesis" are from Morrison (2010). The distinction is quite common in the literature, albeit under different names. See P. L. Quinn's (1974) distinction between a "separability" and a "falsifiability thesis" and Ariew's (1984) distinction between a "non-separability" and a "non-falsifiability thesis."

(PT) *Prediction thesis*: a single theoretical sentence does not imply an observation categorical. Only clusters of theoretical sentences will imply observation categoricals.[10]

(FT) *Falsification thesis*: whenever a predicted observation categorical turns out to be false, one cannot logically determine which theoretical sentence is falsified. Rather, the cluster of theoretical sentences that implied the categorical is falsified as a whole.[11]

Note that FT is a logical consequence of PT: if only a cluster of hypotheses H&A implies the observation categorical O, then one cannot conclude not-H from not-O; by modus tollens, only not-(H&A) follows from the failed prediction. Still, there are scholars who both accept PT and reject FT. They fail to see that FT merely states that there is no *logical* way to decide between alternative revisions, a claim that is perfectly compatible with the claim that there are other reasonable means to decide between competing hypotheses.[12]

Like empiricism, holism is a thesis that is itself supported by empirical findings; Quine believes it to be an empirical fact about scientific practice that scientists have many options to restore a theory's consistency with observation in the light of adverse experience. Second, Quine believes that evidential holism is justified by the complexity of the language we use to formulate scientific theories. Our scientific language is so complicated that it cannot be learned by "continuous derivation" from observation sentences. In consequence, we also cannot follow this process backward and "reduce scientific theory to sheer observation" (NNK, 1975d, 267). Nonholistic languages are possible, according to Quine, but they would never be rich enough to express our best scientific theories, or so he argues in a response to Robert Nozick:

> [Nozick] asks whether a non-Duhemian language would be impossible for us. Let me say that the observation sentences, in my behaviorally defined sense, constitute already a rudimentary language of this kind.

[10] There is one set of trivial exceptions to PT: if one combines all the theoretical sentences that together imply an observation categorical into one long conjunction, this conjunction will imply the categorical by itself as well. See (FME, 1975a, 72; RJV, 1986f, 620).

[11] See (PT, 1990a, 13–14): "the falsity of the observation categorical does not conclusively refute the hypothesis. What it refutes is the conjunction of sentences that was needed to imply the observation categorical. In order to retract that conjunction we do not have to retract the hypothesis in question; we could retract some other sentence of the conjunction instead."

[12] See Sober (1999; 2000) and, for a response, Verhaegh (2017b). Indeed, Quine has often attempted to list the "pragmatic maxims" that describe our revision norms beyond the norms of logic. See (PR, 1955, 247), (WO, 1960, §5), (WB, 1970, ch. 6), (TI, 1990b, 11), and (FSS, 1995a, 49).

> [. . .] But I see no hope of a science comparable in power to our own that would not be subject to holism. (RRN, 1986c, 364)

Holism, in short, is an empirical thesis; it is firmly supported by observations about scientific practice and language learning.[13]

So why does Quine believe that evidential holism supports naturalism? What he seems to have in mind is the following: once we realize, on the basis of PT and FT, that we cannot translate our theoretical terms into epistemologically more basic sensory concepts, we ought to acknowledge that the Cartesian dream of providing an absolute science-independent foundation for our scientific beliefs should be given up. If Quine's ideas about the holistic relation between theory and evidence are correct, in other words, the classical empiricist project of "deducing science from sense data" (EN, 1969b, 84) simply cannot be carried out.

As we have seen in chapter 2, however, Quine's argument is stronger than this; not only does Quine believe that we ought to "despair of being able to define theoretical terms generally in terms of phenomena," he also argues that this project is flawed from the beginning since the sense data the classical empiricists appealed to do not constitute a truly science-independent foundation to start with. Quine, we have seen, argues that "[s]ense data are posits too" (PR, 1955, 252), such that our ideas about sense experience themselves depend on prior scientific theorizing. As a result, even if it *were* possible to translate our theoretical concepts in terms of sense data, such a reduction would not constitute a truly science-*independent* foundation for science.

As a result, the first half of our characterization of Quine's naturalism—NT above—is supported by his holism; Quine rejects a detached, extrascientific perspective on reality because his holistic picture of inquiry leads him to the conclusion that such a perspective simply cannot be had. Next to his rejection of an *a priori* science-independent perspective on the basis of his empiricism, therefore, Quine also dismisses the possibility of an *a posteriori* science-independent perspective; the sense data posited by traditional epistemologists themselves depend on scientific theory.[14]

[13] See also Gibson (1988, pp. 32–34), who dubs these arguments the "scientific practices argument" and the "language learning argument." For a more detailed discussion of Quine's defense of PT and FT, see Verhaegh (2017b).

[14] Or, as he put it in his Oxford Lectures of 1953: "To start from scratch [and] obtain science by pure incontrovertible reason—that's out. The rationalist dream. To start with a *tabula rasa* and fill it in with pure experience by [definition and] logic—that's out. The phenomenalist dream. But something of the original ill-formulated motivation of epistemology does remain valid. The [twofold] root of scientific knowledge: 1) the stimulation of our end organs by rays and molecules [. . .] 2) our inner apparatus [. . .] inherited or acquired, for weaving theories. All this is subject matter for science" (1953–1954, item 3283, my transcription).

Realism: Let us turn to the second source of naturalism Quine identifies in "Five Milestones of Empiricism":

> The other negative source of naturalism is unregenerate realism, the robust state of mind of the natural scientist who has never felt any qualms beyond the negotiable uncertainties internal to science. (FME, 1975a, 72).

Why does Quine cite realism as a source of naturalism? It is my contention that the answer can be found in "The Pragmatists' Place in Empiricism," the paper on which "Five Milestones of Empiricism" is based.[15] In this paper, Quine compares his naturalism with the views of William James, Ferdinand Schiller, and John Dewey. According to Quine, these classical pragmatists "viewed science as a conceptual shorthand for organizing observations" (PPE, 1975b, 33), such that we cannot ascribe reality to our scientific posits and theories. Now, given Quine's ideas about underdetermination, the view that there exist alternative theories that would equally fit our observational evidence, Quine seems committed to such an instrumentalist reading of science as well. After all, his underdetermination thesis seems to imply that "the systematic structure of scientific theory [. . .] is invented rather than discovered, because it is not uniquely determined by the data." Quine, however, argues that he is *not* committed to the instrumentalism of James, Schiller, and Dewey:

> For naturalistic philosophers such as I [. . .] physical objects are real, right down to the most hypothetical of particles, though this recognition of them is subject, like all science, to correction. I can hold this ontological line of naive and unregenerate realism, and at the same time I can hail man as largely the author rather than discoverer of truth. I can hold both lines because scientific truth about physical objects is still the *truth*, for all man's authorship. [. . .] We are always talking within our going system when we attribute truth; we cannot talk otherwise. (PPE, 1975b, 33–34)

Realism, in other words, is an important feature of Quine's naturalism; without it his position would lapse into instrumentalism. After all, instrumentalism is a variant of empiricism too, a variant, moreover, which is perfectly compatible with some varieties of evidential holism.[16]

[15] "Five Milestones" is a shortened version of "The Pragmatists' Place in Empiricism." See (TT, 1981a, 67n).

[16] Quine once believed that his underdetermination thesis was a corollary of evidential holism. In "Indeterminacy of Translation Again" (ITA, 1987b, 345), however, Quine argues that the underdetermination thesis is merely "suggested by" evidential holism. In a letter to Michele Leonelli, Quine explains why: "It had occurred to me only rather lately that the under-determination of science is a

"Five Milestones" provides us with a first characterization of the commitments supporting Quine's naturalism. Both NT and SI are supported by Quine's empiricism, holism, and realism. A question that remains to be answered, however, is how Quine justifies his realism. We have seen that both empiricism and evidential holism are theses which are themselves supported by science. Have we here finally found a science-independent, distinctively philosophical presupposition in Quine's naturalism? I believe not. As the last sentence of the above quote shows, Quine believes that we *cannot but* think about our scientific theories as true; "we are always talking within our going system when we attribute truth; we cannot talk otherwise." This claim, as we shall see, should be taken quite literally: according to Quine, key philosophical concepts (like "truth") are useless when they are divorced from their everyday scientific applications.

4.3. A Quirk of Usage

SI states that the naturalist tentatively accepts our inherited scientific theories and methods. The "tentative" part is important; it expresses Quine's strong commitment to fallibilism—his firm belief that "no statement is immune to revision" (TDE, 1951a, 43). Yet the "inherited" part, often overlooked in discussions of Quine's naturalism, is equally important. In many key passages in his oeuvre, Quine expresses his commitment to the view that in "talking within our going system," we are working from within an "inherited" conceptual scheme:

> The naturalistic philosopher begins his reasoning within the inherited world theory as a going concern. He tentatively believes all of it, but believes also that some identified portions are wrong. He tries to improve, clarify, and understand the system from within. He is the busy sailor adrift on Neurath's boat. (FME, 1975a, 72)

> Analyze theory-building how we will, we all must start in the middle. Our conceptual firsts are middle-sized, middle-distanced objects, and our introduction to them and to everything comes midway in the cultural evolution of the race. In assimilating this cultural fare we are little more aware of a distinction between report and invention, substance and style, cues and conceptualization, than we are of a distinction between the proteins and the carbohydrates of our material intake.

direct corollary of holism. But this is wrong. Holism assures us that any recalcitrant observation can be accommodated in various ways; but these way[s] fare differently in the face of future observations. Conceivably (though improbably) only one theory—*Iddio sa quale*—is compatible with all possible observations" (February 6, 1987, item 637).

> Retrospectively we may distinguish the components of theory-building, as we distinguish the proteins and carbohydrates while subsisting on them. (WO, 1960b, §1)

> Each man is given a scientific heritage plus a continuing barrage of sensory stimulation; and the considerations which guide him in warping this scientific heritage to fit his continuing sensory promptings are, where rational, pragmatic. (TDE, 1951a, 46)

Quine's model of inquiry, in other words, consists of two basic elements: (1) we all start in the middle, assimilating our inherited world theory, and (2) we work from within this inherited system as we go along, modifying the system while relying on its best theories and methods. Quine's claim that our conceptual scheme is "inherited" explains why SI does not require a distinct philosophical justification. It is simply an empirical fact that we all assimilate the "lore of our fathers" (CLT, 1954a, 132) when we grow up.[17]

Prima facie, Quine's model of inquiry seems to be a fairly trivial. Of course we all acquire a certain world theory when we grow up; and of course no one will deny that this theory will play a significant role in our initial inquiries. Quine's picture does not seem to capture a fundamental distinction between realists and instrumentalists or between naturalists and nonnaturalists. Many philosophical perspectives seem to be compatible with the view that the "philosopher begins his reasoning within the inherited world theory as a going concern."

Quine's view, however, is significantly stronger than it might first appear. For Quine believes that his picture of inquiry has far-reaching metaphysical and epistemological consequences. For one thing, Quine's unregenerate realism is supported by this view. When instrumentalists like James, Schiller, and Dewey claim to accept science but, at the same time, regard it "as literally false on

[17] In a 1965 note, Quine calls the idea that we are always working from within an inherited conceptual scheme "involutionism." In the note, which is transcribed in the appendix (A.6), he defines involutionism as the view that "[w]e must work within a growing system to which we are born" (July 21, 1965, item 3182). The note also sheds light on who might have influenced Quine in this respect. For, in the note, Quine attributes the view to Otto Neurath and Karl Popper. Quine's attribution of involutionism to Neurath makes sense given his fondness of the latter's boat metaphor. Quine's attribution of involutionism to Popper is probably prompted by his reading of *Conjectures and Refutations* (1962). See Quine's library, item AC95.Qu441.Zz962p. In the book, Popper argues that "without tradition, knowledge would be impossible. Knowledge cannot start from nothing—from a *tabula rasa*—nor yet from observation. The advance of knowledge consists, mainly, in the modification of earlier knowledge" (1962, p. 28). In any case, Quine's choice of the term "involutionism" makes clear that he has in mind something like SI. To be involuted is to be immanent. Indeed, in his note, Quine also mentions "immanentialism" as an alternative label. In section 7.5, I will analyze Quine's note in more detail.

ontological points" (PPE, 1975b, 35), they presuppose a science-independent notion of "truth"—they are not really working from within.

Quine's argument here can be illustrated using his theory of truth.[18] Quine, as is well known, defends a deflationary theory of truth. According to Quine, our truth predicate is nothing more than "a device of disquotation" (PL, 1970a, 12). To claim that a sentence like "Snow is white" is true is simply to claim that snow is white:

> Where it makes sense to apply 'true' is to a sentence couched in the terms of a given theory and seen from within the theory, complete with its posited reality. [. . .] To say that the statement 'Brutus killed Caesar' is true, or that 'The atomic weight of sodium is 23' is true, is in effect simply to say that Brutus killed Caesar, or that the atomic weight of sodium is 23. (WO, 1960b, 24).[19]

Intuitively, Quine's theory of truth seems misguided; most people (including scientists) seem to presuppose a more transcendental conception of truth. That is, most people seem to be convinced that truth lies beyond their theories and that their beliefs are true or false once and for all, regardless of what they believe. Few people, for example, would accept that our planet started orbiting the sun only a few hundred years ago when Copernicus's hypothesis became generally accepted. Rather, they say that Copernicus discovered something about our solar system that has been true all along:

> [U]sage dictates that when in the course of scientific progress some former tenet comes to be superseded and denied, we do not say that it used to be true but became false. The usage is rather that we thought it was true but it never was. (WDWD, 1999a, 164)

In talking about falsified theories as having been false all along, in other words, we seem to presuppose a notion of truth that does not fulfill Quine's scheme: even when our best scientific theories implied that the earth is the center of the universe, the statement "The sun orbits the earth" failed to be true. Like traditional

[18] Indeed, in the note mentioned in footnote 17, Quine uses his "involutionism" primarily to argue against certain conceptions of truth.

[19] See also (PT, 1990a, 80): "To ascribe truth to the sentence ['Snow is white'] is to ascribe whiteness to snow. [. . .] Ascription of truth just cancels the quotation marks." It should be noted that Quine does not defend a naive form of disquotationalism. According to Quine, the truth paradoxes teach us that "if a language has at its disposal the innocent notations for treating of quoting and appending, and also the notations of elementary logic, then it cannot contain a truth predicate that disquotes all its own eternal sentences"—especially the sentences which themselves contain the predicate "true" (PT, 1990a, 83). Quine adopts Tarski's definition of truth to circumvent the truth paradoxes. See (PL, 1970, ch. 3).

metaphysicians, we seem to presuppose that our ideas about the solar system might be false, even if our best theories imply that they are true. In talking about our beliefs we seem to be presuppose a substantive, transcendental theory of truth; we seem to believe that the question whether or not our theories are *really* true is a legitimate question.[20]

Quine, however, believes that his theory can accommodate this "quirk of usage" (RA, 1994d, 230). For the conclusion that a recently falsified hypothesis has been false all along does not require a transcendental notion of truth. His theory of truth explains this way of talking perfectly, as long as we recognize that this usage itself reflects an element of our inherited world theory. The realism "integral to the semantics of the predicate 'true'" (FSS, 1995a, 67) is just a theoretical choice, a choice guided by the same standards, for example, as the choice to adopt a heliocentric worldview. Just as our belief that the earth orbits the sun is a consequence of our theory about the solar system, our belief that the earth has *always* orbited the sun (regardless of what we believed about it) is a consequence of our theory that the workings of the world around us do not depend on our beliefs about that world.[21] The claim that our best scientific theories might be false, in other words, is itself part of our inherited world theory. Our belief that the world might be different from what we believe it to be (our "unregenerate realism," so to speak) is deeply entrenched, but it is not a belief somehow divorced from our scientific conceptual scheme.[22]

[20] Quine discusses this issue in his (FSS, 1995b; NLOM, 1995c; RE, 1995d; and WDWD, 1999a). A related phenomenon which Quine discusses in some of these papers, is the fact that whenever we contemplate hypotheses on which we will never be able take a stand, we still believe these hypotheses to be either true or false. We will probably never find out, for instance, whether or not there was an even number of grass blades in Boston Common at the inception of 1901. Still we believe that the hypothesis that there was an even number of grass leaves is either true or false. We presuppose that all assertoric sentences are truth-valued, regardless of whether we will ever be able to establish these truth-values, and hence, we seem to presuppose the idea of a transcendental truth. See (WDWD, 1999a, 165).

[21] Cf. Hylton (2007, pp. 277–78).

[22] In a similar way, Quine solves the "grass blades problem" mentioned in footnote 20. According to Quine, we do not need to presuppose the idea of a transcendental truth in order to account for the fact that we believe all assertoric sentences to be truth-valued, even if we will never be able to establish the truth-value of some sentences. For we only need to presuppose the law of the excluded middle to derive from Quine's theory the thesis that all assertoric sentences are either true or false. That is, if we add the law of the excluded middle to a Quine's account of truth, it automatically follows that all assertoric sentences are truth-valued, even if we will never be able to take a stand on some of them:

> [L]et 'p' stand for a sentence to the effect that there were an even number of blades of grass in Boston Common at the inception of 1901. By excluded middle, p or not p; so, by disquotation, 'p' is true or 'Not p' is true. (WDWD, 1999a, 165)

Quine's theory of truth illustrates that his picture of inquiry has strong metaphysical implications. If we take the view that we are "working from within" seriously, even our realism and our ideas about truth will be immanent.[23] Quine's view implies that we ought to stop interpreting key philosophical concepts in a transcendental, extrascientific fashion. According to Quine, "The concept of truth [and, we might add, the concept of reality] belongs to the conceptual apparatus of science on a par with the concepts of existence, matter, body, gravitation, number, neutrino, and chipmunk" (WDWD, 1999a, 165). When the traditional metaphysician asks us about the true nature of reality and when James, Schiller, and Dewey adopt an instrumentalist reading of science, they presuppose extrascientific notions of "truth" and "reality." Quine's view, however, takes seriously the fact that we are working from within and dismisses such extrascientific notions:

> [W]e own and use our beliefs of the moment, even in the midst of philosophizing, until by what is vaguely called scientific method we change them here and there for the better. Within our own total evolving doctrine, we can judge truth as earnestly and absolutely as can be; subject to correction, but that goes without saying.[24] (WO, 1960b, 25)

4.4. Deflation, Deflation, Deflation

Quine's model of inquiry consists of two basic elements: (1) we all start in the middle, assimilating our "inherited world theory"; and (2) we work from within this inherited system as we go along, relying on our best theories and methods. Element (1) explains why SI does not require a distinct philosophical justification. It is simply an empirical fact that all start in the middle. Element (2) explains why we should be unregenerate realists, and hence why Quine's view does not lapse into instrumentalism. Quine's picture of inquiry helps us understand what it means to be committed to SI; if we take seriously the view that we are working

The law of the excluded middle, in turn, is also just another element of our inherited world theory; it is adopted "for the simplicity of theory it affords" (WPB, 1981c, 32). That is, the decision whether or not to adopt the law of the excluded middle is just a revisable theoretical decision, not a transcendental one.

[23] In a letter to Bergström, Quine playfully notes that "truth is eminently immanent" (February 24, 1995, item 86, my transcription).

[24] In a letter to Ernest Sosa, Quine writes: it is "only within science, subject to correction still from within science, that we say what there is and how things are; such is my naturalism. The question whether science is true of reality makes no sense transcendentally; it makes sense only as a repetition, within science, of the very question that science has already been tentatively and fallibly answering with its scientific theory" (September 20, 1983, item 1014).

from within, we have to accept that our notions of "truth" and "reality" should be interpreted immanently as well.[25]

Still, Quine's picture of inquiry does not only have metaphysical consequences, it also has radical implications for our ideas about justification. In "Epistemology Naturalized," as we have seen (chapter 2), Quine notoriously claims that if we abandon Carnap's program of epistemological reduction, epistemology "simply falls into place as a chapter of psychology" (EN, 1969b, 84). This remark has often been interpreted as a rejection of normative epistemology tout court:

> [T]he Cartesian program [is] one possible response to the problem of epistemic justification, the two-part project of identifying the criteria of epistemic justification and determining what beliefs are in fact justified according to those criteria. In urging "naturalized epistemology" on us, Quine is not suggesting that we give up the Cartesian foundationalist solution and explore others within the same framework. [. . .] Quine's proposal is more radical. [. . .] He is asking us to set aside the entire framework of justification-centered epistemology. That is what is new in Quine's proposals. Quine is asking us to put in its place a purely descriptive, causal-nomological science of human cognition.[26] (Kim, 1988, p. 388)

Now, Quine has often rejected this reading of "Epistemology Naturalized." According to Quine, naturalized epistemology is *not* a purely descriptive enterprise:

> It is a mistake to suppose that the naturalizing of epistemology deprives it of normative force. The normative aspect subsists. (1987, item 3182(3), pp. 2–3, my transcription)

> Naturalization of epistemology does not jettison the normative and settle for the indiscriminate description of ongoing procedures. (RMW, 1986g, 664–65)

> I agree with [traditionalists] that repudiation of the Cartesian dream is no minor deviation. But they are wrong in protesting that the normative element, so characteristic of epistemology, goes by the board. (PT, 1990a, 19)

[25] Quine explicitly makes the connection between naturalism, instrumentalism, and the antitranscendentalist argument in a response to Christopher Hookway: "Hookway finds 'Two Dogmas' instrumentalist. I think this is fair, and that it applies to my later work as well. But realism peeps through at the checkpoints, and takes over altogether when we adopt a sternly naturalistic stance and recognize 'real' as itself a term within our scientific theory" (RA, 1994d, 233).

[26] See also, for example, Putnam (1982) and White (1986).

Quine, in other words, explicitly dismisses the interpretation that he replaces normative epistemology with descriptive psychology.

How we *should* think about Quine's normative epistemology becomes clear when we take seriously his picture of inquiry. We have seen that Quine's picture supports his deflationary, immanent conception of truth; in working from within, we also commit ourselves to an immanent notion of "truth." Now, the very same move can be made in epistemology. As some scholars have recognized, Quine also adopts an immanent, *deflationary theory of justification*, i.e. a theory which does not seek a substantive extrascientific explanation for the justification of our statements beyond their being included in or excluded by our inherited world theory.[27]

Although Quine's project has a significant descriptive component, his epistemology remains normative because it appeals to such a deflationary notion of justification—a notion according to which facts about how we *actually* construct theory from evidence coincide with facts about how we *should* do this. When Quine defines naturalism as the view according to which our scientific theories are "not in need of any justification *beyond* observation and the hypothetico-deductive method" (FME, 1975a, 72), in other words, he is not rejecting talk about justification; he is claiming that our best scientific theories are justified *in virtue of* their being based on observation and the hypothetico-deductive method—in virtue of their being our best scientific theories.[28] Our inherited scientific theories are not in need of any extrascientific justification because they are already justified in virtue of their being the best theories at our disposal. Combining his deflationary theories, we can conclude that Quine is not denying that our theories are true and justified; he is rejecting substantive extrascientific standards of reality, truth, and justification.[29]

[27] See, for example, Gregory (2008), Ebbs (2011b), and Sinclair (2014). Ebbs aptly dubs Quine's theory a "minimalist understanding of justification" (2011b, p. 630).

[28] See also Johnsen (2005, p. 88): "so far is [Quine] from proposing to abandon the normative that he is proposing instead to *discover* the norms that govern theorizing by discovering the norms that we conform to in our theorizing. [. . .] What he is [. . .] proposing is to *enlist the aid of psychology* in addressing the burden of epistemology: psychology will identify the norms we adhere to, and philosophy [i.e. his naturalistic picture of inquiry] will tell us that, *by virtue of* their being the ones we adhere to, they are the ones we *are to* adhere to."

[29] A similar response can also be offered against the common misunderstanding that Quine's theory of meaning implies that we cannot use our sentences to make assertions:

> [I]f we are to give up the notion of sentence-synonymy as senseless, we must give up the notion of sentence-significance [. . .] as senseless too. But then perhaps we might as well give up the notion of sense. (Grice and Strawson, 1956, p. 146)

In "Quine Gets the Last Word," Ebbs effectively argues that these and related objections fail to take into account Quine's "resolutely minimalist understand of language use" (2011b, p. 617). Using

4.5. The Bounds of Science

We have seen how Quine's picture of inquiry informs his commitment to SI; if we take seriously the view that we are working from within, we should also adopt deflationary, immanent conceptions of reality, truth, and justification. Before we move on to NT, the second component of Quine's naturalism as specified above, I want to say a bit more about Quine's use of the term "science."

SI states that the naturalist tentatively accepts our inherited *scientific* theories and methods—that "each man is given a *scientific* heritage plus a continuing barrage of sensory stimulation." This terminology, combined with the above-discussed misunderstanding that Quine replaces normative epistemology with descriptive psychology, has often led scholars to conclude that Quinean naturalism is a deeply "scientistic" worldview. Indeed, although naturalism *in general* has become increasingly popular in the past few decades, many contemporary naturalists dismiss *Quine's* variant of the position as overly scientistic.[30] In fact, many contemporary naturalists have urged for a variant of naturalism that is more modest; i.e. a variant of naturalism that can be qualified as "soft," "open minded," "naïve," "harmless," "nonscientistic," "pluralistic," and/or "liberal."[31]

Any such interpretation of SI is unwarranted, however, as Quine always uses the term "science" broadly:

> We have no word with the breadth of *Wissenschaft*, but that is what I have in mind. History is as at home in my naturalism as physics and mathematics.[32] (RGE, 1995e, 34)

the terminology developed in this chapter, we can rephrase Ebbs's argument as the claim that Quine's picture of inquiry also has consequences for his understanding of language use: Quine is not denying that our utterances are meaningful; he is rejecting transcendental standards of significance. See also section 3.5.

[30] See, for example, Lugg (2016, p. 204), who argues that Quine's way of proceeding is "not much in favor nowadays," as even philosophers who, like Quine, "despair of past philosophy," regard his *scientism* "as a step too far" (my emphasis). See also Haack (1993a, 1993b), Siegel (1995), Almeder (1998), Glock (2003), Putnam (2004), and Weir (2014). For a reply to Haack, see Verhaegh (2017c). Some scholars classify Quine's naturalism as "scientistic" without intending to use the term pejoratively. See, for example, Gibson (1988) and Gaudet (2006).

[31] See Strawson (1983), Stroud (1996), Hornsby (1997), Almeder (1998), De Caro and Macarthur (2004b), and Dupré (2004) respectively. The last term ("liberal naturalism") is from Mario De Caro and David Macarthur, who have edited two volumes (2004a, 2010a) devoted to developing this position. They define "liberal naturalism" as a philosophical perspective "that wants to do justice to the range and diversity of the sciences, including the social and human sciences [. . .] and to the plurality of forms of understanding, including the possibility of non-scientific, nonsupernatural forms of understanding" (2010b, p. 9).

[32] See also (FSS, 1995b, 49), (NLOM, 1995c, 462), and (RTE, 1997, 255). In a letter to Christopher Hookway, Quine protests against the narrow conception of science that is ascribed to

Indeed, when Quine was confronted with P. F. Strawson's distinction between "*strict* or *reductive* naturalism" on the one hand, and a more "*catholic* or *liberal* naturalism" on the other (1983, 1), he simply responded by claiming that the difference between the two variants "seems to waver and dissolve" (FHQP 1985b, 208), suggesting that Strawson "wouldn't find the discrepancy" between the two if he had simply adopted "a more liberal or catholic conception of science" (1993, item 2498, 13).

Again, the key to understanding Quine's notion of science is his picture of inquiry. For Quine's view that we all start in the middle entails that there is no strict distinction between science (in the narrow sense) and other types of knowledge; our "inherited world theory" will always be a mixture of science (in the narrow sense), history, superstition, local custom, and common sense. Moreover, Quine does not believe that there is anything special about our scientific knowledge (in the narrow sense); the norms of science are continuous with the methods that guide us in our everyday inquiries:

> [S]cience is [. . .] a continuation of common sense. The scientist is indistinguishable from the common man in his sense of evidence, except that the scientist is more careful. This increased care is not a revision of evidential standards, but only the more patient and systematic collection and use of what anyone would deem to be evidence. If the scientist sometimes overrules something which a superstitious layman might have called evidence, this may simply be because the scientist has other and contrary evidence which, if patiently presented to the layman bit by bit, would be conceded superior. (SLS, 1954b, 233)

Our best scientific theories and methods, in other words, should be merely viewed as a *refinement* of the theories and norms we all use in our everyday inquiries.[33]

SI, in other words, should not be viewed as a scientistic thesis if "science" is interpreted narrowly. In response to the question as to wherein our perspective should be immanent, Quine's answer, of course, is "science." Quine's

him in Hookway (1988): "All these passages ascribe a far narrower conception of science to me than I hold" (May 31, 1988, item 529). Quine's commitment to a broad notion of science is certainly not new: already in "Two Dogmas of Empiricism," Quine equated "total science" with "[t]he totality of our so-called knowledge or beliefs, from the most casual matters of geography and history to the profoundest laws of atomic physics or even of pure mathematics and logic" (TDE, 1951a, 42).

[33] See also, for example, Quine's (WO, 1960, 3), "science is self-conscious common sense"; his (NK, 1969c, 129), "[Science] [. . .] differs from common sense only in degree of methodological sophistication"; and his (WIB, 1984d, 310), "science is refined common sense."

notion of science, however, is quite broad. The view that Quine's naturalism is scientistic, in other words, presents his position the wrong way around. It is only *after* adopting a broadly science-immanent perspective that Quine, in improving and clarifying his system from within, starts making choices that some contemporary philosophers have argued to be unduly restrictive.[34] Quine's naturalism *itself* is not a scientistic thesis; his ideas about the *ultima facie* bounds of science are established from within as well (Ricketts, 1982).

4.6. Immanence and Transcendence

Let me sum up what we have established thus far. Quine's naturalism can be characterized by NT and SI. SI is based on his view that we are working from within our inherited world theory. If we take seriously Quine's picture of inquiry, we see both why SI should not be viewed as a scientistic thesis and why the naturalist is committed to deflationary theories of truth, reality, and justification.

Let us now turn to NT, the second component of Quine's naturalism. What exactly does Quine reject when he dismisses transcendental perspectives? We have already seen some examples of perspectives Quine rejects: he dismisses the phenomenologist's idea of a science-independent justification of science and the traditional metaphysician's attempt to examine what reality is *really* like. But what do such transcendental perspectives have in common? And what is Quine's argument for dismissing them?

Let me start by noting that both in everyday life and in the sciences, we often attempt to transcend our subjective points of view by stepping back from our personal perspectives and by developing a more "objective" picture about what is the case and about what only appears to be the case. If I note that the piece of paper I am writing on appears to be orange, for example, I might examine the paper in different lighting conditions to examine whether the paper is really orange or whether it only appears to be orange due to the light bulb in the lamp on my desk. Likewise, I can try to establish an even more "objective" perspective on the color of the paper by using a spectrophotometer. In this way, we rise above our everyday beliefs by inquiring more and more carefully into the "reality" of these beliefs. We transcend our initial subjective point of view by adopting a perspective which encompasses the relation between our initial subjective outlook and the world we inhabit. Of course, this is not the type of transcendence that

[34] E.g. his physicalism (FM, 1977; OW, 1978b) or his choice to regard certain components of our inherited world theory as "Grade B idiom" (RTD, 1968g, 305).

Quine is objecting to, as it is the very business of scientists to inquire into the reality of our beliefs in this way.[35]

A second type of transcendence that is harmless, in Quine's view, is what he calls "semantic ascent," the technique of switching from talk about things to talk about words and from talk about our beliefs and theories to talk about systems of sentences (FSS, 1995a, 92). Semantic ascent is useful according to Quine because in discussing our theories with disagreeing opponents, we do not want these very theories to influence our judgment. Hence, we step back from our theories, regard them as systems of sentences, and compare the properties they have vis-à-vis our shared evidence and scientific norms:

> Philosophical issues often challenge the basic structure of our system of the world. When this happens, we cannot easily dissociate ourselves from our system so as to think about our opponent's alternative. The basic structure of our system inheres in our very way of thinking. Thus the discussion can degenerate into question-begging, each party stubbornly reiterating his own basic principles, the very principles that are at issue. But we can rise above this predicament by talking about our theories as systems of sentences; talking about the sentences instead of just stubbornly asserting them. [. . .] We can find a common ground on which to join issues instead of begging the question.[36] (IQ, 1978a, 15–16)

Like ordinary inquiry into the reality of our beliefs, semantic ascent is not problematic. For in order to accept the technique as an acceptable form of transcendence, we only need to admit that it is possible to develop a metalanguage in which we can talk about our first-order theories.

Both types of harmless transcendence (ordinary scientific inquiry and semantic ascent) can be repeated. If we can shift from talk about theories to talk about systems of sentences, thus ascending to a metalanguage, we can also ascend to a *meta*metalanguage and shift from talk about systems of sentences to talk about systems of sentences about systems of sentences. The same holds for scientific inquiry into the question what is the case and what only appears to be the case. If we can transcend our initial subjective perspective by "improving, clarifying, and understanding our system from within," we can also transcend this improved perspective by further inquiry.

In theory, therefore, we are able to take an infinite number of steps beyond our initial subjective points of view—beyond our inherited world theory. For every

[35] In (QSM, 1988, 25), Quine approvingly quotes David Justice's claim that "[i]n the scientific enterprise man transcends his language just as he transcends his untutored sense impression."

[36] See also (WO, 1960, §56).

new perspective can be transcended by either approach. Still, our perspective of the world will always remain limited in some sense. For every new perspective will remain an essentially human perspective, bounded by our modes of perception and our existing conceptual scheme. In asking about the true nature of reality, therefore, the traditional philosopher asks for something more; she asks us whether the objects in the world are colored in themselves, and she asks us whether our beliefs are true independent of any language. In general, the traditional philosopher is not satisfied with the (theoretical) possibility of an infinite number of perspectives, every new perspective still more "objective" than the previous one. Rather, by inquiring into the true nature of reality, the traditional philosopher attempts to develop an absolutely objective perspective on the nature of things. In Quine's words, she asks us to "detach ourselves" from our conceptual scheme, to inquire "into the absolute correctness of [this] conceptual scheme as a mirror of reality," and thus to compare our beliefs "objectively with an unconceptualized reality" (IOH, 1950b, 79). Let us adopt Thomas Nagel's term here and call this the "view from nowhere" (1986). When Quine argues that "there is no [. . .] cosmic exile" (WO, 1960b, 275), he is arguing against the existence of a such view from nowhere—he is arguing that we ought to dismiss "the question whether or in how far our science measure up to the *Ding an sich*" (TTPT, 1981b, 22).

Still, this is not all there is to NT. If it were, virtually no contemporary philosopher would disagree with Quine's rejection of transcendental perspectives.[37] Quine rejects a view that is far less controversial; he dismisses the view that it is possible to abandon our current conceptual scheme in order to adopt a new and radically *different* one. That is, he is not only rejecting the view from *nowhere*, he is also rejecting the possibility that we can transcend our current conceptual scheme by moving to a novel and distinct alternative—he is also rejecting the possibility of what might be called "a view from *somewhere else*." Indeed, in dismissing the traditional epistemologist's idea of a science-independent sensory language, Quine is not rejecting the view from nowhere. The sense-datum language, after all, is still a language. Nor are sense data independent of our modes of perception. Rather, as we have seen in chapter 2, Quine dismisses phenomenalism because it purports to offer a *science-independent* perspective—because it purports to offer a view from somewhere else. Likewise, in rejecting an instrumentalist reading of science, Quine is not rejecting the view from nowhere either. After all, James, Schiller, and Dewey would have despised the idea

[37] See Siegel (1995, p. 50): if transcendentalism is defined as the aim to "transcend *all* conceptual schemes at once, and see things from some scheme-neutral perspective," then "[p]hilosophers generally—traditionalists and naturalists alike—reject [. . .] transcendentalism."

as much as Quine. Rather, Quine is dismissing their instrumentalism because it relies on a science-independent notion of "truth."

NT, in other words, is also grounded in Quine's picture of inquiry. If we take seriously Quine's commitment that we are always working from within, then not only are we to dismiss the view from nowhere, we are to dismiss all views from anywhere else. It is never possible to completely abandon one's inherited world theory and to adopt another or start anew. Any change of theory, any improvement made, will be a gradual internal modification, and rely to a certain extent on what was already there. We can only use what Quine has called "Neurath's plank-by-plank methodology."[38]

4.7. Artificial Languages

One might object that the position I am ascribing to Quine is too strong. It cannot be the case that Quine is dismissing *anything* outside our evolving world theory because he always allowed the construction of artificial languages—the construction of alternative conceptual schemes. As a logician, for example, Quine worked within a diverse range of logical frameworks,[39] and in *Set Theory and Its Logic* (1963), Quine compares multiple axiomatic set theories.

Yet Quine's perspective on artificial languages exactly illustrates my interpretation of NT. For Quine always insisted that artificial notations are only useful if they are grounded in natural language—if we can assess them from within. Indeed, as I will argue in chapter 6, disagreement about artificial languages goes to the very core of the Carnap-Quine debate about analyticity. For as Quine explains in a letter to Alonzo Church, Carnap assumes that we can simply define terms like analyticity and synonymy by constructing semantical rules, whereas Quine believes that serious artificial notations depend on what we already have:

> [M]y attitude toward 'formal' languages is very different from Carnap's. Serious artificial notations, e.g. in mathematics or in your logic or mine,

[38] See (RS1, 1968h, 316). The fact that Quine often uses Neurath's metaphor to illustrate his view—e.g. in (IOH, 1950b, 79), (OME, 1952, 223–25), (PR, 1955, 253), (WO, 1960, 3, 124, 210), (NK, 1969c, 126–27), and (FME, 1975a, 72)—might suggest that Neurath's writings played an important role in Quine's development. Quine himself, however, claims there was no such influence. In a letter to Dirk Koppelberg, who compared Quine's and Neurath's views in *Die Aufhebung der analytischen Philosophie* (1987), Quine writes: "my reading of my predecessors has been very sporadic and inadequate. I was aware superficially of my affinity with Neurath, as you know, and I am glad now to see the degree to it and the detail. I was not appreciably influenced by him at the time; I had to grow into the point of view on my own, away from Carnap" (April 18, 1986, item 601, my transcription).

[39] For an overview, see Decock (2004, §3).

> I consider supplementary but integral parts of natural language. [. . .] Thus it is that I would consider an empirical criterion [. . .] a solution of the problem of synonymy in general. And thus it is also that [. . .] I am unmoved by constructions by Carnap in terms of so-called 'semantical rules of a language'.[40] (August 14, 1943, item 224)

What Quine means when he talks about empirical criteria for synonymy, he explains in a letter to Carnap:

> The point can be made more clearly if we begin, by way of analogy with the notion of sentence. The empirical linguist who goes into the field to study and formulate a language unrelated to any languages hitherto formulated has a working idea, however vague [. . .] of *sentence* in general. [. . .] Now when for theoretical discussion we specify an artificial language as object, we [also] specify [. . .] the class of expressions which are to be regarded as sentences for this language. The idea of "sentence" is the same in both cases; only the languages are different. [. . .] It is only thus that we understand what is intended, what is coming, when you tell us: "the following are to be the sentences of my new language." [. . .] Now my view on the notion of "analytic" [. . .] is similar to my view on the notion of sentence, in this respect: It is only by having some general, pragmatically grounded, essentially behavioristic explanation of what it means in general to say that a given sound- or script-pattern is analytic for a given individual, that we can understand what is intended when you tell us (via semantical rules, say) "the following are to be analytic in my new language". Otherwise your specification of what is analytic for a given language dangles in midair. (QCC, May 10, 1943, 337–38)

Artificial languages, in other words, are no exception to Quine's rejection of perspectives transcending our inherited world theory; Quine believes that artificial notations are only useful if they are grounded in natural language—if we can assess them from within.

4.8. Conclusion

I have argued that, at the most fundamental level, to be a Quinean naturalist is to "work from within." Quine's model of inquiry consists of two basic elements: (1)

[40] I thank Warren Goldfarb and Thomas Ricketts for their suggestion to study the Quine-Church correspondence.

we all start in the middle, assimilating our "inherited world theory"; and (2) we work from within this inherited system as we go along, relying on our best theories and methods. Prima facie, Quine's view about the nature of inquiry appears relatively innocent. I have argued, however, that Quine's picture has far-reaching philosophical consequences: it leads him to adopt deflationary theories of truth, reality, and justification, as well as novel views about the possibility of transcendence and the nature of artificial languages. Accordingly, Quine's picture of inquiry helps us to understand why he adopts NT and SI, the two core components of Quinean naturalism.

PART II

DEVELOPMENT

5

Sign and Object

5.1. Introduction

To be a Quinean naturalist is to work from within. But how did Quine himself arrive at the conclusion that there is no distinctively philosophical perspective on reality? In the preceding chapters, I have examined Quine's naturalism, its scope, and its far-reaching consequences for philosophy. In the remainder of this book, I will study the *genesis*, the *development*, and the *reception* of Quine's ideas. Just as Quine himself aims to understand the justificatory structure of science by adopting a genetic approach, I examine the structure of Quine's naturalism by studying the evolution of his metaphilosophical thinking.

As of yet, little attention has been paid to the evolution of Quine's naturalism. Although historians in recent years have started to become increasingly interested in postwar analytic philosophy, little work has been devoted to the steps Quine took in developing his views.[1] Given that Quine did not endorse a fully naturalistic perspective until the mid-1950s, this fact seems particularly surprising. In the early stages of his career—the 1930s, the 1940s, and the early 1950s—Quine, as we shall see, still defended views that he would later dismiss. Indeed, Quine himself, as we have seen, noted that he became "more consciously and explicitly naturalistic" only in the 1950s; that is, "in the ten years between 'Two Dogmas' and *Word and Object*" (TDR, 1991b, 398).

A similar conclusion can be drawn about the *reception* of Quine's views. Although it is evident that Quine's views about the nature of philosophy have strongly influenced the course of the analytic tradition, surprisingly little is known about the way in which Quine's work was actually received. Where the

[1] See, for example, Janssen-Lauret and Kemp (2016, p. 2), who argue that "[m]ore historical awareness of Quine is urgently needed." For examples of some pivotal first steps toward the study of Quine's development, see Creath (1987), Dreben (1992), Hylton (2001), Decock (2002a), Ben-Menahem (2005), Isaac (2005), Mancosu (2008), Ebbs (2011a), Føllesdal (2011), Frost-Arnold (2011), Lugg (2012), Murphey (2012), Sinclair (2012), and Morris (2015).

reception study is a common methodological tool in most subfields of the history of philosophy, it is an instrument that is still rarely used by historians of *analytic* philosophy.[2]

Fortunately, an astonishing number of documents relating to the development and the reception of Quine's views—notes, academic and editorial correspondence, drafts, lectures, teaching materials, and grant proposals—has been saved and stored at the W. V. Quine Papers at Houghton Library. These documents provide a unique opportunity to minutely reconstruct the evolution and reception of Quine's ideas in a crucial stage of his development, i.e. the period in which he grew from a respected logician to a world-leading philosopher.

In the upcoming chapters, I discuss these documents and argue that the development of Quine's naturalism can be divided into three mutually reinforcing (and largely parallel) subdevelopments: (1) the evolution of his views about metaphysics and epistemology (chapter 5); (2) his growing discontent with the analytic-synthetic distinction and his adoption of a holistic explanation of logical and mathematical knowledge (chapter 6); and (3) his growing recognition that his changing views about epistemology, metaphysics, and the analytic-synthetic distinction imply that there is no strict distinction between science and philosophy (chapter 7).

In this first chapter, I discuss a series of notes, letters, and manuscripts from the 1940s and early 1950s, documents which show that Quine in the early 1940s started working on his first philosophical monograph—a book entitled *Sign and Object*. The Houghton archives contain several sets of autograph and typewritten notes from the 1941–1946 period connected to what Quine called his "book on ontology."[3]

Thus far, Quine's first philosophical book project has gone completely unnoticed in the literature.[4] In this chapter, I aim to contribute to our understanding of the development of Quine's philosophy by reconstructing his position in *Sign*

[2] See Floyd's call for such reception studies in her survey article "Recent themes in the history of early analytic philosophy" (2009, pp. 199–200): "the decade during which the middle Quine and the later Wittgenstein were initially being put to work in academic philosophy, and the story of their receptions [. . .] is a fascinating one. The history of the interpretations of such philosophers is itself part of the wider story, and very much a part of where philosophy is now."

[3] Quine to Nelson Goodman (December 19, 1944, item 420). Quine also mentions his book project in letters to, among others, Alonzo Church (February 15, 1942, item 570), D. C. Williams (April 7, 1942, item 1221), Morton Wurtele (October 10, 1944, item 1244), B. F. Skinner (February 23, 1945, item 1001), Rudolf Carnap (May 10, 1945), and Paul Buck (November 30, 1945, item 473). In his letter to Carnap, Quine even speculates that the book might become his "*magnum opus*."

[4] The only exception is Murphey (2012), who briefly discusses some of Quine's notes from 1943 and 1944 (pp. 53–55). Murphey's account is unreliable, however, because he examines only a very small fragment of the documents related to *Sign and Object*.

and Object as well as the developments that led him to give up on the project in 1946. I argue that although *Sign and Object* was naturalistic in many respects, Quine encountered significant problems in formulating his position, problems which shed new light on his early position, his evolving views about Carnap's philosophy, and the place of "Two Dogmas of Empiricism" in Quine's oeuvre.

This chapter is structured as follows. After a sketch of the philosophical background of Quine's book project (section 5.2), I introduce his method of inquiry in *Sign and Object* and examine a note on "the nature of metaphysical judgments," showing that Quine still defends a mixture of Carnapian and mature views in 1941 (sections 5.3–5.4). Furthermore, I argue that Quine's perspective on metaphysics can be traced back to 1937, when he for the first time expresses doubts about Carnap's quasi-syntactical approach to metaphysics (section 5.5). Next, I show that Quine adopts a fully naturalistic metaphysics in 1944 (section 5.6) but postpones his book project because he fails to come up with satisfying solutions to problems in epistemology and the philosophy of language (section 5.7). I argue that although "Two Dogmas of Empiricism" is widely considered to be a breakthrough in his thinking, Quine himself was largely dissatisfied with the paper because he felt that he had failed to provide solutions to these problems (section 5.8). In the final sections, I reconstruct the development of Quine's alternative to Carnap's picture by studying, among others, a series of unpublished papers, letters, and lectures on epistemology—papers which led Quine to reboot his book project, eventually culminating in *Word and Object* in 1960 (sections 5.9–5.10).

5.2. Philosophical Background

Although Quine primarily thought of himself as a logician in the 1930s, philosophy was never far away; while his early publications are outwardly technical, many of them implicitly intend to clarify problems in ontology, semantics, and the philosophy of mathematics. Looking back on his earliest work, Quine argues that even his dissertation ("The Logic of Sequences: A Generalization of *Principia Mathematica*") was already "philosophical in conception." Actively concerned with the paradoxes in logic and set theory, Quine aimed to transform logic and mathematics as simply and as economically as possible, thereby aspiring "to comprehend the foundations of logic and mathematics and hence of the abstract structure of all science" (TML, 1985a, 85).[5]

[5] See Morris (2015) for discussion. Morris argues that Quine, in transforming logic and mathematics without worrying about prior philosophical commitments, is already committed to the naturalistic ideal that it is the philosopher's task to "scrutinize and improve the system from within," only

More overtly philosophical was Quine's active concern with the ontological presuppositions of logical frameworks. Not only was he already interested in the relation between logic and ontology in his Warsaw discussions with Stanisław Leśniewski in May 1933,[6] also in one of his very first publications—"Ontological Remarks on the Propositional Calculus"—Quine adopts an explicitly ontological perspective on logical systems, arguing that we ought to eliminate propositions because

> the whole notion of sentences as names is superfluous and figures only as a source of illusory problems. Without altering the theory of deduction internally, we can so reconstrue it as to sweep away such fictive considerations; we have merely to interpret the theory as a formal grammar for the manipulation of sentences, and to abandon the view that sentences are names. (ORPC, 1934c, 59)

A few years later, at the 1939 International Congress for the Unity of Science, Quine first presented his criterion of ontological commitment.

A related philosophical development can be distilled from Quine's early work on nominalism. Although Quine claims to have already "felt a nominalist's discontent with classes" in 1933 (IA, 1986a, 14), his early work on the matter seems to be at least mildly skeptical.[7] This attitude seems to change somewhat, however, when Quine—in the 1940–41 Harvard Logic Group meetings with Carnap, Tarski, Goodman, and Russell—discusses the prospects of a finitist-nominalist language of science.[8] Although Quine, in one of the meetings, presents a paper in which he argues that he does not "insist on eliminating classes or other unthingly objects" because it "is not clear that the unthingly can be eliminated without losing science" (December 20, 1940, item 2954), he

"appealing to coherence and simplicity" (WO, 1960, 276). My account in this chapter is to a large extent complementary to Morris's; where Morris argues that "much of the basic philosophical outlook that led to [Quine's] later philosophy is present as early as [his] 1932 dissertation" (2015, p. 152), I am interested here in the subsequent development of this basic philosophical outlook.

[6] See (TML, 1985a, 104): "With Leśniewski I would argue far into the night, trying to convince him that his system of logic did not avoid, as he supposed, the assuming of abstract objects. Ontology was much on my mind." See also Quine's 1933 letter to C. I. Lewis, in which Quine summarizes his academic adventures in Europe as a Sheldon Travel Fellow and mentions that "[s]o far as my more specific interests in logic are concerned, Prof. Leśniewski's work looms largest." He continues the letter with a discussion of the ontological presuppositions of the latter's systems "Protothetic" and "Ontology" (1933, item 643, my transcription).

[7] In an unpublished 1937 lecture titled "Nominalism," for example, Quine argues that the position is "incompatible with ordinary logic and mathematics" (October 25, 1937, item 2969, my transcription). For a reconstruction, see Mancosu (2008, §1).

[8] Carnap's notes of these discussions are published and extensively introduced in Frost-Arnold (2013).

seems to intensify his attempts to "set up a nominalistic language in which all of natural science can be expressed" (DE, 1939b, 708) in the years immediately following the meetings—attempts that eventually culminate in his joint paper "Steps toward a Constructive Nominalism" (Goodman and Quine, 1947).[9]

A final development that seems to have contributed to Quine's increased philosophical aspirations in the early 1940s is his growing discontent with Carnap's views on language and ontology. Whereas Quine seems to have defended a largely Carnapian perspective on the nature of philosophy in the earliest stages of his career,[10] he grew more skeptical of the latter's views during the second half of the 1930s, as is evinced by his explicit opposition to Carnap during the Harvard Logic Group meetings.[11] As we shall see in section 5.5, Quine's opposition to Carnap's views on *metaphysics* can also be traced back to this period.

Considering these developments—his first formulation of the criterion of ontological commitment, his increased concern with the status of abstract objects, and his growing discontent with Carnap's philosophical system—it is not surprising that Quine, after finishing his third logical (text)book in 1940,[12] decided to devote a more substantial part of his time to philosophy. Indeed, in a grant application from January 1941, Quine applies for a small sum of money for secretarial assistance in connection with, among others, investigations into "the philosophical presuppositions of science," a proposal that is heavily influenced by the discussions in the Harvard Logic Group. For Quine proposes to investigate

> (a) The ontological presuppositions of mathematics and natural science. [. . .] Can science be formulated as not to presuppose universals? (b) The problem of infinity, and its connection (via the Skolem-Löwenheim Theorem) with the problem of universals. (c) The epistemological difference, if any, between mathematics and natural science. (January 9, 1941, item 475)

[9] See Gosselin (1990) and Mancosu (2008, §3) for a discussion. A transcription of Quine's paper for the Harvard Logic Group meeting is included in the appendix (A.2).

[10] Indeed, after rereading his "Lectures on Carnap" (1934b) in 1986, Quine writes: "I am startled to see how abjectly I was hawing Carnap's line in my 1934 lectures" (April 10, 1986, item 270). In "Two Dogmas in Retrospect," Quine calls his lectures on Carnap "abjectly sequacious" (1991b, 391).

[11] In these meetings Quine objected, among others, to Carnap's appeal to the analytic-synthetic distinction as well as to his semantic turn in *Introduction to Semantics* (1942), a manuscript the group used as a basis for discussion. See (TDR, 1991b, 392) and, for a discussion, Mancosu (2005) and Frost-Arnold (2011). Some scholars believe that Quine is also arguing against Carnap's view in "Truth by Convention," arguably Quine's most influential philosophical contribution in the 1930s. For an alternative interpretation of "Truth by Convention," see Ebbs (2011) and Morris (forthcoming).

[12] I.e. *Elementary Logic* (1941a). Quine's first two logical (text)books were *A System of Logistic* (1934a) and *Mathematical Logic* (1940).

The earliest evidence of Quine's plan to write his first philosophical *monograph* in connection with these (and other) investigations is from November 1941, when he wrote a four-page untitled book outline.[13]

5.3. Starting at the Middle

Quine's book outline shows that he planned to write a book that is fairly comprehensive in philosophical coverage; the seven chapters he envisions are titled (1) A tentative ontology, (2) Time and tense, (3) Sign and object, (4) Extensionality, (5) Meaning, (6) The place of epistemology, and (7) Nominalism and empiricism (November 1941, item 3169, my transcription). From a developmental perspective, one of the most interesting parts is Quine's outline of the opening chapter, which begins as follows:

> Starting at the middle: an ontology that may want supplementation *or* diminution. (Concrete) *things* in some sense or other (spatio-temporal regions, or quanta of action, or bundles of quanta of action, etc., for the present left undecided), and all attributes and relations of such things, and all attributes and relations of the thus supplemented totality, and so on.
>
> Just *what* attributes and relations *are* there? Well, common-sense partial answer is that every condition we can formulate (everything of the form of a statement, but with a variable in place of one or more signs of entities) determines an *attribute* of just the entities fulfilling that condition, a *relation* of just the *n*-ads of entities fulfilling that condition. (Roughly: whatever we say about an object attributes an attribute to that object; and so on). This is the *principle of abstraction* (*comprehension*).
>
> [. . .] But not so fast. Even principle of abstraction untenable! (Paradoxes.) There must be some conditions with no attributes or relations corresponding. A modicum of nominalism is thrust upon us *despite* our having begun with ever so willing a Platonism. (November 1941, item 3169, my transcription)

Although Quine's argument is still rather sketchy, several elements are interesting from a contemporary perspective. First and foremost, it should be noted that Quine uses a largely naturalistic method of inquiry. Like the first section of *Word and Object*, in which he argues that "we all must start

[13] The outline is simply titled "Book." A few months later, however, Quine has adopted the title *Sign and Object*. In a letter to his Harvard colleague D. C. Williams, for example, Quine writes that he has "begun to outline" a "little book, *Sign and Object*" (April 7, 1942, item 1221).

in the middle" and opens with the "familiar desk" that "manifests its presence by resisting my pressures and by deflecting light to my eyes" (1960, §1), Quine, in his outline, is also "starting at the middle." Moreover, Quine's ideas about what constitutes "the middle" resemble his later views as well. Just as in *Word and Object*, "the middle" seems to refer to a mixture of science and common sense. After all, Quine starts with both a scientific ontology ("quanta of action") and with what he himself calls the common-sense view about attributes.[14]

Second, the model of inquiry presupposed in these opening remarks is consistent with his mature view that although we *start* at the middle, we are trying to *improve* this system from within. For although Quine seems to want to settle on a nominalistically acceptable ontology,[15] he *justifies* this decision by appealing to the paradoxes he encounters while working from within his common-sense idea about attributes and relations. Although Quine is starting at the middle, in other words, he is emphasizing that this starting point "may want supplementation or diminution."

5.4. The Nature of Metaphysical Judgments

Despite the apparent continuity between his early and his mature views about the nature of inquiry, however, there is evidence that Quine did not or could not yet completely practice what he preached. In this section and the next, I argue that Quine had not yet completely worked out how to naturalize his ideas about metaphysics.

In 1939, as we have seen, Quine first set out his criterion of ontological commitment in a paper for the Harvard conference on the unity of science.[16] Despite

[14] The suggestion that Quine was already committed to a gradualist picture concerning the distinction between science and common sense is confirmed by a paper he read one year earlier for the Harvard Logic Group. In this lecture, Quine initially distinguishes between a world of "ordinary things" like "tables [and] chairs" and a scientific "super-world" consisting of "electrons [and] atoms," but later concludes that "the dichotomy [. . .] between science and common sense is surely false" (December 20, 1940, item 2954). Again, a transcription of this paper is included in the appendix (A.2).

[15] I use the qualifier "seems" here because it is far from clear what are Quine's actual aims in the later chapters of the projected book. The outline of the seventh chapter of *Sign and Object* ("Nominalism and empiricism"), for instance, is too sketchy to draw any reliable conclusion about Quine's aims.

[16] A six-page abstract of Quine's conference paper "A Logistical Approach to the Ontological Problem" was to be published in the *Journal of Unified Science* but never appeared due to the German invasion of the Netherlands. See (LAOP, 1939a, 64). Quine did publish part of the paper as "Designation and Existence" in the *Journal of Philosophy* (1939b).

Quine's ontological breakthrough, however, his paper is neutral with respect to the traditional questions of metaphysics. After all, it solely deals with the question what entities we are committed to *if* we accept a particular language.[17] It remains an open question what language we are to adopt—a question whose answer depends on one's views about metaphysics.[18]

The details of Quine's early views about metaphysics become clearer when we turn to his 1941 book outline. For, in the outline, Quine also explicitly states his broader, more general, perspective on the nature of metaphysical existence claims. On the very first page, even before the above-discussed sketch of the first chapter, Quine argues that metaphysical statements are factual but that they cannot be evaluated rationally:

> Nature of metaphysical judgments:
>
> As factual as any; not quasi-syntactical. But can't be criticized in the way other factual judgments can, because they concern the very fdtns. of the conceptual scheme relative to which we criticize other judgments. Bootstraps (Kant).
>
> What *can* be criticized is an *analogue*, viz: A *believes* such & such metaphysical statement. Here we *do* examine A's *language* to decide; insofar, Carnap right. But the analogue is not *equivalent*.
>
> Of course there *is* also met. that is *bad* because of *vagueness*; and to be expected, since, as observed, not subject to control. (November 1941, item 3169, my transcription)

There is much to unpack in this short passage. A first thing that is remarkable from a contemporary perspective is that Quine distinguishes between metaphysical statements (first paragraph) and ontological commitments (second paragraph) and explicitly claims that the two are "not equivalent." Where the mature Quine collapses the distinction between ordinary and metaphysical existence claims by arguing that we can answer the question whether or not *x*'s exist by answering the question whether or not our best scientific theories commit us to *x*'s, here he seems to claim that although questions about ontological commitments

[17] As Quine would put it almost a decade later: "We look to bound variables in connection with ontology not in order to know what there is, but in order to know what a given remark or doctrine, ours or someone else's, *says* there is; and this much is quite properly a problem involving language. But what there is is another question" (OWTI, 1948, 15–16). See section 3.4.

[18] That is, the traditional metaphysician might argue that we should adopt a language that can be used to mirror what the world is *really* like, whereas Carnap would answer that we are free to adopt any language we like as long as its rules are clearly specified. The mature Quine, of course, rejects Carnap's distinction between questions of fact and questions of language choice. See chapter 3. In the early 1940s, however, Quine had not yet conclusively rejected Carnap's distinction.

can serve as an "*analogue*" of metaphysical judgments, they are not completely alike.[19]

Second, even if we accept Quine's distinction between ontological commitments and metaphysical existence claims, his ideas *about* these metaphysical judgments are remarkable from a contemporary perspective as well; for although Quine holds that metaphysical statements are factual, he also maintains that they cannot be criticized, i.e. that they are "not subject to control." Where Quine's distinction between metaphysical and ordinary existence claims in the first and second paragraphs still seems to be in Carnapian spirit—corresponding, somewhat anachronistically, to the latter's external and internal questions (1950) respectively—he proposes an important emendation here. He agrees with Carnap that metaphysical judgments cannot be rationally evaluated, but he does not want to succumb to the latter's suggestion that these questions are without cognitive content. Indeed, he implicitly refers to Carnap in claiming that metaphysical existence claims are not "quasi-syntactical" but "as factual as any."[20]

The mature Quine of course provides a plausible explanation as to why metaphysical judgments are factual. For in naturalizing metaphysics, metaphysical questions are equated with regular queries about ontological commitments, i.e. questions that can only be distinguished from scientific queries in "breadth of categories." Where the natural scientist deals with "wombats and unicorns" and simply assumes a realm of physical objects, the metaphysician/ontologist scrutinizes the question whether we ought to accept "the realm of physical objects itself" (WO, 1960, 275). This is not the solution Quine adopts in his outline, however. After all, Quine's mature position implies that metaphysical judgments *can* be criticized: just as we can rationally evaluate the natural

[19] One might object that Quine, in considering the "analogue" question whether or not someone "believes such & such metaphysical statement," is not really talking about ontological commitments. After all, if one is *committed* to a certain entity by means of Quine's criterion, then one does not necessarily have to *believe* the corresponding metaphysical statement. A nominalist might believe that there are no abstract entities, for example, while being unknowingly committed to such entities because of the sentences she utters. The second sentence of the second paragraph shows that this objection is misguided, however: Quine's idea that we can "decide" whether or not A believes a certain metaphysical statement by examining "A's language" shows that he equates the two. From Quine's behavioristic perspective this makes perfect sense: to know what A believes *is* to know what sentences he will be disposed to assent to in appropriate circumstances.

[20] In *The Logical Syntax of Language* (1934), Carnap construes metaphysical statements as quasi-syntactical statements, meaning that although these statements appear to be about extralinguistic objects, they are in fact about linguistic expressions. Or, as Quine summarized it himself: "in the quasi-syntactic idiom we appear to be talking about certain nonlinguistic objects, when all we *need* be talking about is the sign or signs themselves which are used for denoting those objects" (LC, 1934b, 98).

scientist's claim that there are no unicorns, we can evaluate the ontologist's claim there are no abstract objects.

The first outline of *Sign and Object*, in other words, still presents a mixture of Carnapian and mature views. Quine argues that metaphysical statements are factual (and not quasi-syntactical), but he still agrees with Carnap that they cannot be criticized, showing that he had not yet completely dismissed Carnap's argument for rejecting metaphysical statements. The result is an apparently inconsistent mixture of ideas about the relation between metaphysics and ontology. For it is unclear how Quine, given his strong empiricist commitments, would have explained that metaphysical claims are factual even if they are not "subject to control."

5.5. A Pragmatic Interpretation of Positivism

Although Quine's ideas about metaphysics in *Sign and Object* are remarkable from a contemporary perspective, it is interesting to see that he was already trying to move away from Carnap. This suggests that there might be an explanation for Quine's view if we consider the question *why* he rejected Carnap's position.

It is my contention that there are two fundamental reasons for Quine's discontent with Carnap's quasi-syntactical approach, and I believe that both sources of dissatisfaction can be traced back to at least 1937.[21] First, Quine seems to believe that Carnap's position is too easy—that he is dodging his ontological commitments.[22] Just as Quine in 1932 had tried to convince Leśniewski that "his system of logic did not avoid, as he supposed, the assuming of abstract objects" (TML, 1985a, 104), he also felt that Carnap could not avoid being committed to abstract objects, even if he was dismissing metaphysical claims as quasi-syntactical. In his 1937 lecture "Nominalism," for example, Quine describes Carnap as a "second-order nominalist" and subtly accuses him of taking the easy way out:

> *Carnap* has been considered a nominalist, though he doesn't incline to the sacrifices of the intuitionist. But he is a nominalist only in a very

[21] I use the qualifier "at least" because it is unclear if Quine ever *completely* accepted Carnap's perspective on metaphysics. The last sentences of his "abjectly sequacious" 1934 lectures on Carnap, for example, are somewhat ambiguous: "Views will differ as to the success of Carnap's total thesis that all philosophy is syntax. Carnap has made a very strong case for this thesis; but it must be admitted that there are difficulties to be ironed out. [. . .] He has in any case shown conclusively that the bulk of what we relegate to philosophy can be handled rigorously and clearly within syntax. Carnap himself recognizes that this accomplishment stands *independently* of the thesis that *no* meaningful metaphysics remains beyond syntax. Whether or not he has really slain the metaphysical wolf, at least he has shown us how to keep him from our door" (LC, 1934b, 102–3).

[22] See section 3.3.

> different sense. A nominalist of second intention: Doesn't nominalize universals, but nominalizes the problem *of* universals.
>
> [. . .]
>
> Nominalism in its ordinary form has perhaps two purposes: 1st) to avoid metaphysical questions [. . .] 2nd) To provide for reduction of all statements to statements ultimately about tangible things, matters of fact. This by way of keeping our feet on the ground—avoiding empty theorizing. Carnap's 2[n]d-degree nominalism succeeds in the 1st respect [. . .] But it accomplishes nothing in the 2nd respect. (October 25, 1937, item 2969, my transcription)

According to Quine, in other words, Carnap's second-order approach serves as some sort of ontological free pass; it allows us to talk about abstract entities without worrying about our ontological commitments. Or, as he would put it even less sympathetically almost ten years later:

> Logical positivists didn't take Platonistic implications of math. seriously. They were opportunistic: argued against metaphysics as non-empirical because they didn't like it; admitted math. unquestionably because they like it.[23] (ca. 1947, item 3266, my transcription)

The second problem Quine has with Carnap's account concerns the details of the latter's quasi-syntactical treatment of metaphysical existence claims. Quine not only dismisses *the effects* of the latter's quasi-syntactical approach, claiming that it allows the positivist to dodge her ontological commitments, but also rejects the very idea that metaphysics ought to be dismissed *as meaningless*. In a 1937 note—titled "A Pragmatic Interpretation of Positivism"— Quine rejects Carnap's appeal to a strict criterion of significance in dismissing metaphysics.

> Meaninglessness must be abandoned as meaningless—at least insofar as it might be used against metaphysics. Even supposing we would make sense ultimately of an operational criterion this would rule out all the non-intuitionistic part of math. also.[24] (April 2, 1937, item 3169, my transcription)

[23] This quote is from Quine's notes for his philosophy of language course (Philosophy 148) in 1947.

[24] See section 3.5. It is perhaps no coincidence that Quine started to reflect on criteria of significance in 1937. For this is also the year in which Quine first used set theory instead of type theory in his attempts to dissolve Russell's paradox. One of Quine's first set-theoretic papers was "On Cantor's Theorem" (1937). In a letter to Church, Quine explains that the "motive of the system is [. . .] the elimination of the hierarchy of types, with the complex notion of meaningfulness which it imports" (May 10, 1937, item 224).

Quine, in sum, has two problems with Carnap's perspective on ontology and metaphysics. He believes (1) that Carnap's solution is too easy, as it allows us to talk about abstract objects without worrying about ontological commitments; and (2) that we should get rid of the idea that metaphysics is meaningless.

From Quine's mature naturalistic perspective, there is a uniform solution to both problems. If one collapses the distinction between metaphysics and ontology, arguing that we can answer the question whether or not universals exist by answering the question whether or not our best scientific theories commit us to universals, then one will have replaced both Carnap's second-order nominalism and his claim that metaphysics is meaningless. The 1941 outline shows that Quine has not yet drawn this conclusion, however. Instead of collapsing the distinction between ontological commitment and metaphysics, he maintains the gap by arguing that the two are "not equivalent" and by sticking to Carnap's conclusion that metaphysics cannot be criticized.[25]

5.6. The Philosopher's Task

Between September 1942 and November 1945, Quine served as a full lieutenant (and later as a lieutenant commander) in the navy. As a result, Quine's research time was severely limited for more than three years. In a 1942 letter to Alonzo Church, the main purpose of which was to announce that he was resigning as the consulting editor of the *Journal of Symbolic Logic*, Quine writes that "every bit of time that I can spare from teaching duties must go into war work" and that he has dropped his "research for an indefinite period":

> Before Dec. 7 [the date of the attack on Pearl Harbor], I was outlining a little philosophico-logical book to be entitled *Sign and Object*. All I can hope to do now is synopsize the main ideas, eventually, in a brief article; and I'm not going to think about this, even, till six months or so hence, when teaching and war work have resolved their mutual conflict somehow.[26] (February 15, 1942, item 570)

[25] Quine seems to draw a highly similar conclusion in 1937. In the remainder of the abovementioned note, Quine writes: "Present this from standpoint: What do we *do* & *say* actually? Rather than from an ontological standpoint. Avoid ontology—hence avoid any position regarding whether or not metaphysical & math'l words *denote*" (April 2, 1937, item 3169, my transcription). It appears that Quine here, as in his 1941 note, maintains the gap between ontological commitment and metaphysics: Quine seems interested in "what we *do* & *say* actually" but separates this from metaphysical and ontological questions of existence.

[26] When Quine wrote the letter, he was still combining his war work with teaching. This changed when he reported in Washington for duty on October 7, 1942. See (TML, 1985a, 181).

Despite the fact that Quine had to relegate all his research activities to spare time, there are strong indications that he never stopped working on *Sign and Object* during his navy years, for the majority of Quine's notes for the book project are from this period.[27]

For our present purposes, one of the most interesting documents is a series of notes from October and November 1944, in which Quine explicitly reflects on the relation between philosophy and science and seems to have resolved his earlier, theoretically awkward, view that metaphysical statements are factual but unchallengeable.[28] Whereas the above-discussed outline from 1941 at best indicates that Quine's ideas about the nature of inquiry were already fairly naturalistic, these notes leave no doubts about the maturity of his ideas in the early 1940s. For not only does Quine implicitly adopt a "starting at the middle" approach, as in his earlier outline, he also explicitly adopts a naturalistic view in answering the question of how we ought to conceive of the relation between science and philosophy. Comparing the philosopher's task with the aims of science and mathematics, Quine writes:

> The philosopher's task differs from that of the natural scientist or mathematician, no less conspicuously than the tasks of these latter two differ from each other. The natural scientist and the mathematician both operate within an antecedently accepted conceptual scheme but their methods differ in this way: the mathematician reaches conclusions by tracing out the implications exclusively of the conceptual scheme itself, whereas the natural scientist gleans supplementary data of what happens around him. The philosopher, finally, unlike these others, focuses his scrutiny on the conceptual scheme itself. Here is the task of making things explicit that had been tacit, and precise that had been vague; of uncovering and resolving the paradoxes, smoothing out the kinks, lopping off the vestigial growths, clearing the ontological slums to change the figure.[29] (November 5, 1944, item 3181, my transcription)

[27] The first dated note from this period is written on January 30, 1943; the last dated note is written on April 29, 1945.

[28] A transcription of this set of notes is included in the appendix (A3). One of these notes, a second book outline, is titled "Sign and Object; or, The Semantics of Being," showing that Quine wrote these notes with the book project in the back of his mind (October 4, 1944, item 3169, my transcription). Indeed, six days after drafting this second outline, Quine writes Mort Wurtele that he is working on "a projected book, more philosophical than my past ones, bearing the tentative title 'Sign and Object'" (October, 10, 1944, item 1244, my transcription).

[29] Cf. (WO, 1960, §56): "What distinguishes between the ontological philosopher's concern and all this is only breadth of categories [. . .] it is scrutiny of this uncritical acceptance of the realm of

Quine's perspective on the philosopher's task appears to be well developed. Still, the above paragraph could be interpreted as compatible with Quine's 1941 perspective on metaphysics. After all, the idea that the philosopher "focuses his scrutiny on the conceptual scheme itself" might lead one to suspect that Quine still believes that the philosopher's statements "can't be criticized in the way other factual judgments can" (November 1941, item 3169, my transcription). The next paragraph of the 1944 note, however, rules out this interpretation, thereby showing that Quine had changed his mind:

> It is understandable, then, that the philosopher should seek a transcendental vantage point, outside the world that imprisons natural scientist[s] and mathematician[s]. He would make himself independent of the conceptual scheme which it is his task to study and revise. "Give me που στω [a place to stand]" Archimedes said, "and I will move the world." However there is no such cosmic exile. The philosopher cannot study and revise the fundamental conceptual scheme of science and common sense, without having meanwhile some conceptual scheme, whether the same or another no less in need of philosophical scrutiny, in which to work. The philosopher is in the position rather, as Neurath says, "of a mariner who must rebuild his ship on the open sea." (November 5, 1944, item 3181, my transcription)

The similarity with his mature position is remarkable.[30] Whereas Quine in his early notes seems to allow factual but unchallengeable metaphysical judgments concerned with "the very fdtns. of the conceptual scheme relative to which we criticize other judgments," he seems to be fully committed to a naturalized metaphysics in 1944.[31] Indeed, in his second book outline from October 1944, Quine's opening chapter is titled "The Reality of the External World." In this chapter, Quine considers the question of whether "physics might be said to be

physical objects itself, or of classes, etc., that devolves upon ontology. Here is the task of making explicit what had been tacit, and precise what had been vague; of exposing and resolving paradoxes, smoothing kinks, lopping off vestigial growths, clearing ontological slums."

[30] Parts of this passage appear verbatim in *Word and Object*, published sixteen years (!) later. See (WO, 1960, 275–276, my emphasis): "The philosopher's task differs from the others', then, in detail; but in no such drastic way as those suppose who imagine for the philosopher a vantage point outside the conceptual scheme that he takes in charge. *There is no such cosmic exile. He cannot study and revise the fundamental conceptual scheme of science and common sense without having some conceptual scheme, whether the same or another no less in need of philosophical scrutiny, in which to work.*"

[31] At best, one can detect a difference in emphasis. Where the fully naturalistic Quine is prone to focus on the continuity between science and philosophy, Quine in these early notes is more inclined to emphasize their distinctness. For although they are both working from within, the scientist and the philosopher do not yet seem to be concerned with the same project. Rather, Quine believes that the

concerned with exploring the nature of reality." Instead of contemplating the relation between ontological commitments and metaphysical judgments, Quine worries about challenges that might undercut his naturalistic worldview. In the end, Quine dissolves these challenges and settles on a realistic position ("I *am* being a realist") (October 4, 1944, item 3181, my transcription).[32]

5.7. Two Problems

Despite these similarities between *Sign and Object* and *Word and Object*, Quine was dissatisfied with the progress he was making. In a 1944 letter to Goodman, written shortly after he had drafted his second, more detailed, outline (October 4, 1944, item 3169, my transcription), Quine complains about the book's progress:

> In the matter of logic and philosophy, I'm at more of a standstill than I have been for half a generation. Still dickering with the introduction of a book on ontology.[33] (December 19, 1944, item 420)

It is probably because of this reason that Quine decided to postpone his work on *Sign and Object* in 1946 and to let his philosophical ideas simmer for a few years in his new course on the philosophy of language.[34]

For the purposes of this chapter, it is important to see why Quine was dissatisfied with the progress he was making. It is my contention that there are two fundamental reasons as to why Quine struggled to complete *Sign and Object*. First, Quine had yet to develop a comprehensive view about language, meaning, and the nature of logical and mathematical knowledge. Quine had been dissatisfied with Carnap's analytic-synthetic distinction for years, but he had not

ontological question is "broad enough to allow both philosopher[s] and scientist[s] to move about in it without treading on each other's toes" (November 5, 1944, item 3181). See also chapter 7.

[32] See appendix section A.3. This changed perspective is present in all notes Quine wrote in 1944. In the few notes in which he still uses the terms "ontology" and "metaphysics," he seems to use them interchangeably.

[33] See also the letters Quine wrote to Charles Morris ("I've made miserably little progress," February 9, 1945, item 741) and B. F. Skinner: "I've been wanting to write a book of philosophical and semantic character, but I don't seem to make any headway now. I'm not in the expansive mood that makes for philosophical creativity" (February 23, 1945, item 1001).

[34] In a project proposal for the Rockefeller Foundation, Quine writes: "I have a considerable sheaf of rough fragments [. . .] written over recent years with the idea in mind of a book of primarily philosophical and semantical complexion. These things have not yet shaped up into anything organic, but my thought along these lines has led me to offer, for the coming spring, a course in the Philosophy of Language. Accordingly [this project] will consist immediately in the planning of that course; the book is the ulterior purpose" (July 9, 1946, item 921).

been able to formulate a satisfying alternative. Although he had succeeded in defining analyticity in terms of synonymy in 1943,[35] he really struggled to find a satisfying behavioristic explication of synonymy. Indeed, one of the most detailed notes from Quine's navy years is a document entitled "Foundations of a Linguistic Theory of Meaning" (August 1943, item 3169), in which he examines and dismisses some candidate explications of synonymy. The problem was that even if Quine *did* decide to abandon the whole project of appealing to an analytic-synthetic distinction to account "for the meaningfulness of logical and mathematical truth" (RGH, 1986b, 206–7), he had "no suggestion of a bright replacement" (TDR, 1991, 393). Indeed, in a 1948 letter to Hugh Miller, Quine sums up his predicament pretty well:

> I am with you in questioning the currently popular boundary between analytic and synthetic. I feel, indeed, that the distinction means virtually nothing, pending the devising of some behavioristic criterion such as no semanticist to date have given us an inkling of. But, for the same reason, I don't know what it would mean to say, with you, that arithmetic is *not* analytic.[36] (May 31, 1948, item 724, my transcription)

A second reason for Quine's dissatisfaction with his philosophical progress is the fact that he had not yet been able to develop a satisfying epistemology. Although he had accepted, as we have seen, an unregenerate realism (and a naturalized metaphysics) in his notes from October and November 1944, he continuously struggled with the phenomenalist objection that primary sense experiences are "more real" than tables, chairs, atoms, and electrons:

> There is a sense in which physics might be said to be concerned with explaining the nature of reality. And who contests this? Primarily the Idealist. [. . .] The Idealist would take the perceptions etc. rather as the basic reality, and derive things as constructions, logical constructs (Russell). The study of how to make these constructions is Epistemology. And things are composed not of atoms but of perceptions, sense qualia etc. (October 4, 1944, item 3169, my transcription)

[35] See, for example, Quine's letter to Carnap from January 5, 1943: "It is clearer, I think, to shortcut the question of definitions in connection with the relation between the analytic and the logically true, and to speak directly, rather, of the relation of *synonymity* or sameness of meaning. Given this notion, along with that of logical truth, we can explain analyticity as follows: a statement is analytic if it can be turned into a logical truth by putting synonyms for synonyms" (QCC, 297).

[36] Quine's evolving views on the analytic-synthetic distinction will be discussed in detail in chapter 6.

Where Quine would later dismiss phenomenalism by arguing that sense data are scientific posits and therefore not in any way more fundamental than, for example, triggerings of sense receptors, he had yet to develop a satisfying response to phenomenalism in the early 1940s.[37]

This "problem of epistemic priority" is one of the most significant problems Quine grappled with in the 1940s. There are two distinct, though related, explanations for Quine's preoccupation with the problem. First and foremost, Quine's two philosophical heroes, Rudolf Carnap and Bertrand Russell, had both defended a phenomenalist perspective at one point or another. Quine was an admirer of Carnap's *Aufbau* and had heard Russell defend phenomenalism in his 1940 William James lectures, "An Inquiry into Meaning and Truth," at Harvard.[38]

A second explanation for Quine's struggle with epistemological priority is the fact that the issue was debated in the 1940–1941 Harvard Logic Group.[39] In his notes of these meetings, Carnap writes: "We have not agreed among ourselves whether it is better to begin with thing-predicates or sense-data-predicates. For the first: I and Tarski; Hempel follows Popper. For the second: Goodman and Quine" (June 18, 1941).[40]

[37] Quine did *try* to answer the phenomenalist's challenge: in some notes Quine argues that he can just ignore the phenomenalistic objection because epistemology is "irrelevant" to his ontological project (March 19, 1944, item 3181, my transcription); in some notes he flirts with a proto-version of his mature position that "[p]erceptions are themselves states of physical objects" (January 30, 1943, item 3169, my transcription); and in "On What There Is," Quine argues that "the obvious counsel is tolerance" such that there are multiple conceptual schemes: e.g. phenomenalistic and physicalistic ones (OWTI, 1948, 19). It seems, however, that Quine did not believe these solutions to be satisfactory, as they never reappeared in later papers. For a somewhat different account, see Murphey (2012, pp. 54–55). Murphey's account, however, is based on a very small fragment of the documents related to *Sign and Object.*

[38] Again, it should be noted that Quine's reading of Carnap and Russell is not uncontroversial. See section 1.4.

[39] There is some evidence that Quine was already interested in the issue when he was still a student. In one of his graduate papers, "Concepts and Working Hypotheses," Quine writes that "no analysis of a given experience can yield any other experience which is, in any full sense, the 'bare datum' of the form of experience; any such analysis is, rather, merely a further interpretation" (March 10, 1931, item 3225).

[40] See Frost-Arnold (2013, pp. 189–90). Although the issue does not show up in Carnap's notes before June 18, phenomenalism also seems to have been discussed in an earlier meeting. On May 22, 1941, for example, Goodman writes Quine:

> That was a good defense yesterday of the epistemological approach. [. . .] It will be interesting to see what happens Monday. Carnap's resistance may have softened a little as a result of being shown that his argument that phenomenal sentences are incomplete presupposes the physicalistic basis he uses it to defend. I hope you are as successful at the department meeting [. . .] as you were in getting phenomenalism another hearing in the group. (May 22, 1941, item 420)

In sum, although Quine had developed a naturalistic picture of inquiry in *Sign and Object*, he started to view the book as "a distant objective" (May 16, 1948, item 921) because he was confronted with two theoretical challenges: (1) he saw no plausible alternative to Carnap's analytic-synthetic distinction, and (2) he had not been able to develop a satisfying epistemology. Indeed, when Quine, in 1946, was asked to suggest topics for a Rockefeller conference on the most urgent questions in philosophy ("the work which philosophy in the United States now has to do"), he replied by listing the problem of "cognitive meaning" and the problem of "epistemic priority" as two of his main concerns:

> Clarification of the notion of cognitive meaning, or of the relation of cognitive synonymy of phrases, is needed in order to make sense of the distinction between analytic and synthetic judgments. This distinction has played a central rôle in philosophy from Kant to Carnap, and is of vital importance to the philosophy of mathematics; but there has been little recognition of the fact that the distinction in its general form is undefined pending a definition of cognitive synonymy. (My own view is that the latter definition should be couched in terms of observable linguistic behavior. I have found no satisfactory formulation).
>
> Clarification of the notion of epistemological priority is needed in order to know what the task of epistemology (as distinct e.g. of psychology) is; for, epistemological priority is the direction in which epistemological reduction of knowledge to more fundamental or immediate knowledge seeks to progress. (October 20, 1946, item 921)

5.8. "Two Dogmas"

In January 1950, Max Black invited Quine to read a paper for the APA Eastern Division meeting in Toronto with the specific request to survey "what questions and issues remain still to be settled in the light of the programs and achievements of the previous half century's work" (January 17, 1950, item 31). Quine hesitated ("I am not at my best in historical surveys") but accepted the invitation after Black clarified that the program committee was specifically interested in "an analysis of the achievements to date and an evaluation of the questions which remain in need of further investigation" (January 20 and February 6, 1950, item 31).

The paper Quine read in Toronto, "Two Dogmas of Empiricism," was immediately recognized to be of major importance. Within six months after the APA

See also "Logic, Math., Science" (December 20, 1940, item 2954), transcribed in the appendix (A.2).

meeting, three events were organized to discuss the paper. In February 1950, Chicago's Department of Philosophy organized a debate between Carnap and Quine; in May 1950, there was a symposium on the paper at Stanford; and in the same month the Institute for Unified Science sponsored two sessions with papers on "Two Dogmas" (June 18, 1951, item 1200).[41] Moreover, Quine received dozens of letters with questions, comments, and requests for copies in the year after he first read the paper.[42]

Despite this widespread recognition that he had written an important paper, Quine felt that the massive attention for "Two Dogmas" was largely undeserved. In response to a letter by Paul Weiss, for example, Quine writes:

> [M]y rather tentative negative strictures on analytic-synthetic have had plenty of attention, disproportionate attention. [. . .] I might feel differently if the doctrine concerned were a positive philosophy. But what is it? (a) The observation that the analytic-synthetic distinction has never been adequately def'ned, though all too widely taken for granted. (b) The tentative conjecture that epistemology might develop more fruitfully under some very different sort of conceptualization, which I do not provide. (c) The suggestion that the analytic-synthetic idea is engendered by an untenably reductionistic phenomenalism. (June 18, 1951, item 1200, my transcription)

Quine, in other words, felt that his paper's main contribution was negative and that he had failed to develop a satisfying alternative.[43]

Whether or not one agrees with his assessment of "Two Dogmas," Quine's dissatisfaction with the paper is perhaps less of a surprise if we reconsider the problems that led him to postpone *Sign and Object* in the mid-1940s. Quine's

[41] It should be noted, however, that Carnap and Quine also debated their views on ontology at the Chicago seminar. See (CVO, 1951c).

[42] Quine ran out of reprints of his article, which appeared in the *Philosophical Review* in January 1950, within a few months. A reissue of the article in mimeograph form, produced for the Stanford symposium, was also exhausted in a year (June 5, 1952, item 1263). Looking back on this period, Quine writes: "It is remarkable that my most contested and anthologized paper was an assignment. The response was quick and startling" (TDR, 1991b, 393).

[43] See also Quine's paper "The Present State of Empiricism" (May 26, 1951, item 3015, my transcription): "I do not flatter myself that ["Two Dogmas"] contributes a new idea to philosophy. The paper is negative: an expression of distrust of two doctrines"; and Quine's letter to Joseph T. Clark from April 17, 1951: "I [. . .] feel much less content at criticism than at construction. This is why the ideas of 'Two dogmas,' reiterated for years in my course on Philosophy of Language and in private disputation, were so slow in getting into print. Correspondingly I find that particular lecture course a continuing burden; whereas I would feel much enthusiasm for it if I were to hit upon a substantial and acceptable constructive theory of knowledge and language" (April 17, 1951, item 231). A transcription of "The Present State of Empiricism" is included in the appendix (A.5).

first problem was to find a satisfying alternative to Carnap's analytic-synthetic distinction. Now, although Quine, in "Two Dogmas," abandons the dogma that "there is a fundamental cleavage between truths which are *analytic* [. . .] and truths which are *synthetic* (1951a, 20), the alternative picture he outlines in the famous sixth section of the paper is sketchy at best. Even in the paragraph in which he tries to clarify what he means with "without metaphor," his definition does not seem to satisfy his own strict behavioristic standards of clarity:

> For vividness I have been speaking in terms of varying distances from a sensory periphery. Let me try now to clarify this notion without metaphor. Certain statements [. . .] seem peculiarly germane to sense experience. [. . .] Such statements, especially germane to particular experiences, I picture as near the periphery. But in this relation of "germaneness" I envisage nothing more than a loose association reflecting the relative likelihood, in practice, of our choosing one statement rather than another for revision in the event of recalcitrant experience. (TDE, 1951a, 43)

Although we tend to read Quine's mature epistemology (including his ideas about observation sentences, language learning, and the nature of scientific theories) into statements like these, the picture outlined here is still very sketchy. Indeed, in the last paragraph of the above-cited letter to Weiss, Quine summarizes the sixth section by claiming that it only contains "the tentative conjecture that epistemology might develop more fruitfully under some very different sort of conceptualization" and, most importantly, that he *does not provide any such conceptualization himself.* When we reconsider the core of Quine's *negative* argument in "Two Dogmas"—the claim that the empiricists had failed to come up with a strict, empirically satisfying definition of analyticity—it becomes clear that he is simply using the very same standards to evaluate his own proposal. Quine was dissatisfied with "Two Dogmas" because he did not practice what he preached; if Carnap's epistemology was in need of empirical clarification, so was his own holistic alternative. Indeed, in response to a critical letter by Joseph Clark, Quine admits that "there is [. . .] much more to be said" about notions like "a 'convenient conceptual scheme' and a 'recalcitrant experience', and much that I am not yet able to say" (April 17, 1951, item 231).[44]

[44] Even after Quine had fully developed his mature epistemology in *Word and Object,* he kept dismissing the last section of "Two Dogmas" as "very brief" and "metaphorical," arguing that "it is a waste of time" to further debate the view he had sketched in 1951 (Quine to Schwartzmann, November 21, 1968, item 958). Quine's last word on "Two Dogmas" is from 1991, when he published "Two Dogmas in Retrospect." Yet even forty years after he first presented the paper in Toronto, Quine still seems to believe that the paper did not make a positive contribution: he recalls that he "had not thought to look on [his] strictures over analyticity as the stuff of revolution. It was mere criticism, a

The sketchiness of Quine's positive account becomes even clearer if we reconsider the second problem discussed in section 5.7, i.e. the problem of how to formulate a satisfying response to epistemological phenomenalism. Where the last section of "Two Dogmas" at the very least contains Quine's first step toward the solution of his problem with the analytic-synthetic distinction, it does not address his problem with phenomenalism at all.[45] If anything, Quine seems to have adopted a phenomenalistic picture of epistemology himself in explicating the experiential boundaries of his metaphorical "man-made fabric" in terms of "sense data" (TDE, 1951a, 42–44).

Quine, in other words, appears to presuppose a holistic variant of epistemological phenomenalism in "Two Dogmas": there are pure uninterpreted sense data and there is a conceptual scheme which is to account for those raw experiences as simply and as effectively as possible.[46] Quine's picture, in sum, still seems miles away from a naturalized epistemology in which all talk of science-independent sense data is abandoned and replaced with talk about the physical stimulation of sensory receptors. Indeed, in response to questions from Percy Bridgman and Henry Margenau, Quine explains that it is his aim to "give to the conceptual scheme everything *except* the raw confirmatory experiences, & to find the external purpose of the conceptual scheme in those experiences" (May 1951, item 3015, my transcription and emphasis).[47]

negative point with no suggestion of a bright replacement" (TDR, 1991b, 267). See also Verhaegh (2018a).

[45] Quine does mention that he rejects "reductionistic phenomenalism" in the above-discussed letter to Paul Weiss, but there he is referring to what is called "radical reductionism" in "Two Dogmas"; i.e. the idea that all statements are individually reducible to experience. In rejecting "reductionistic phenomenalism" in "Two Dogmas," in other words, he is rejecting reductionism, not phenomenalism.

[46] P. F. Strawson also mentions this problem in his review of *From a Logical Point of View*. In discussing "Two Dogmas," Strawson remarks that although Quine rejects phenomenalistic reductionism, his "own conception of our 'experiences,' as set over against and impinging upon, the unified body of our belief, seems to be [. . .] slightly naïve" and "predominantly phenomenalistic" (1955, p. 232). It should be noted, however, that Quine, at the time, incorporated his views about epistemology in a pluralistic picture of conceptual schemes. As in "On What There Is," Quine still believed that the phenomenalistic conceptual scheme is one among many. In a 1951 letter to James B. Conant, for example, Quine writes: "The philosopher who epistemologizes backward to sense data [. . .] is fashioning a conceptual scheme just as the physicist does; but a different one, for a different subject. [. . .] He agrees that there are electrons and tables and chairs and other people, and that the electrons and other elementary particles are 'fundamental' in the physical sense. [. . .] But sense data are 'fundamental' in the epistemological sense" (April 30, 1951, item 254).

[47] See appendix section A.5. Furthermore, in response to Paul Weiss's explicit question whether experience itself is a posit, Quine answers that epistemologically "everything is a posit except the flux of raw experience" (April 10, 1951, item 1200).

5.9. Language and Knowledge

Despite the substantial gap between the phenomenalism in "Two Dogmas" and his mature naturalized epistemology, Quine solved his epistemological problems only one year later. In October 1952, Quine read a paper entitled "The Place of a Theory of Evidence" at Yale. The first half of the lecture discusses Quine's familiar idea that, from an epistemological point of view, both everyday and scientific objects are myths, the function of which is to anticipate "immediate subjective experience" (October 7, 1952, item 3011, p. 15). In the second half of the paper, however, Quine develops a new argument. After concluding that even memories of past sense data do not qualify as epistemically "pure," such that only *present* sense data are available as an epistemological foundation, Quine argues:

> [L]et us take account of the ridiculousness of our position. [. . .] The pursuit of hard data has proved itself, at this point at least, to be self-defeating. [. . .] What has been said just now against [. . .] memory applies in some degree to the stream of sensory experience generally. It would be increasingly apparent from the findings of the Gestalt psychologists, if it were not quite apparent from everyday experience, that our selective awareness of present sensory surfaces is a function of [. . .] past conceptualizations. [. . .] it is not an instructive over-simplification but a basic falsification, to represent cognition as a discernment of regularities in an unadulterated stream of experience. Better to conceive of the stream itself as polluted, at each succeeding point of its course, by every prior cognition. [. . .] We would do well to recognize that in seeking to isolate sense data we are not plumbing the depths of reality. (October 7, 1952, item 3011, pp. 17–19)

Quine, in other words, rejects the view that sense data are epistemological starting points because they are themselves scientific posits.[48] And he immediately draws the naturalistic conclusion that if there is no distinctively epistemological perspective to be had, we can view "epistemology as an empirical science" and replace talk about sense data with talk about "the barrage of physical stimuli to which [a man's] end organs are exposed" (pp. 23–25).

Quine, in sum, had solved his problem with phenomenalism by recognizing that sense data do not offer an extrascientific perspective, an idea that perfectly matches the general picture of inquiry he had already sketched in his notes for

[48] Quine's first publishes (an even stronger version of) this conclusion in "On Mental Entities" (OME, 1952, 225): "the notion of pure sense datum is a pretty tenuous abstraction, a good deal more conjectural than the notion of an external object."

Sign and Object. Interestingly, Quine's breakthrough in epistemology also leads him to solve his first problem. After all, his new conception of "epistemology as an empirical science" provides him with the opportunity to develop an alternative to Carnap's epistemology, i.e. a positive story about scientific, logical, and mathematical knowledge that does not rely on an analytic-synthetic distinction. Indeed, only a few months after his Yale lecture, Quine first expresses his mature view that "[w]e can still study the ways of knowing" by studying "the learning of language" and "the acquisition of scientific concepts"; i.e. by studying "the relation of sensory stimulation to the production of scientific hypotheses by people" (April 9, 1953, item 3158).

Having solved both problems that led him to shelve *Sign and Object* in 1946, it is perhaps no surprise that Quine considered breathing new life into his plan to write a philosophical monograph. In March 1952, Quine requests a small grant from the Harvard Foundation for Advanced Study and Research for secretarial assistance for a "book on semantics and theory of knowledge," a book that "has been gradually evolving in connection with the course in Philosophy of Language which I have given several times in the past few years" (March 4, 1952, item 475). Two months later, Quine writes Roman Jakobson—the editor of the Studies in Communication series at MIT Press—that he is thinking about a book entitled *Language and Knowledge*: "For years my thought has been evolving in the direction of such a book; and I have looked upon various of my articles, as well as my course on philosophy of language, as steps toward it" (May 18, 1952, item 1488).[49]

5.10. Conclusion

In the end, it would take Quine seven more years to complete his first philosophical monograph. During these years, as we shall see in chapter 7, Quine was primarily concerned with the development of his genetic account; his study of "the relation of sensory stimulation to the production of scientific hypotheses by people" (April 9, 1953, item 3158). Where Quine's first published attempt

[49] This would not be the last time that Quine changed his working title. Quine settled on the title *Word and Object* fairly late, viz. somewhere between March and May 1959 (item 1488). Besides *Sign and Object* (1941) and *Language and Knowledge* (1952), there is a long list of titles Quine considered, including, among others, *Language and Theory* (June 22, 1955, item 545), *Words and Things* (December 3, 1956, item 205), *Things and Words* (July 15, 1957, item 1423), *Terms and Objects* (November 6, 1958, item 1005), *Words and the World, Language and the World,* and *Meaning and Reference* (March 11, 1959, item 1488). Note that *Language and Knowledge* is the only title that contains a distinctively epistemological term, which further indicates that Quine's epistemological breakthrough played an important role in his decision to reboot his book project.

to develop such a genetic account—"The Scope and Language of Science" (1954b)—still presents a sketchy story that focuses solely on ostension and internal similarity standards,[50] *Word and Object* provides a detailed and complex account.

In this chapter, I have aimed to contribute to our understanding of the evolution of Quine's thought between 1940, when he first set out to write a philosophical monograph, and the early 1950s, when he first felt ready to complete this project. *Sign and Object,* I have argued, sheds new light on the evolution of Quine's ideas. The notes, letters, and manuscripts related to the project reveal that although Quine's views were already surprisingly naturalistic in the early 1940s, two problems prevented him from developing a comprehensively naturalistic view on philosophical inquiry—problems he could only resolve in the early 1950s. Moreover, *Sign and Object* shows that Quine's development should not be understood solely in terms of his struggle with Carnap's views about analyticity, eventually culminating in "Two Dogmas of Empiricism." Both Carnap's views on metaphysics and (Quine's interpretation of) his *Aufbau* played a very substantial role in Quine's development as well. *Word and Object* is one of the most influential works in the history of analytic philosophy because it offers a novel and comprehensively naturalistic perspective on language, metaphysics, and epistemology. *Sign and Object* unearths the steps Quine took in developing this perspective.

[50] In his autobiography, Quine writes that "The Scope and Language of Science" functioned "as a prospectus for [. . .] *Word and Object*" (TML, 1985a, 258).

6

Analytic and Synthetic

6.1. Introduction

In "Two Dogmas of Empiricism," Quine offers two arguments against the analytic-synthetic distinction. The first argument is pretty straightforward: there appears to be no behavioristically acceptable definition of analyticity. Every definition Quine assesses either depends on a term that itself lacks an adequate definition (e.g. synonymy or necessity) or is unacceptably restricted (e.g. because it only defines "analyticity" for a single language). The second argument is connected to what has been called the "dogma of reductionism." As Quine views the matter, Carnap requires an analytic-synthetic distinction in order to solve an age-old problem for empiricism, viz. the problem of how to explain logical and mathematical knowledge. In arguing that our logical laws and mathematical theorems are analytic, Carnap is able to maintain that these statements are meaningful while remaining faithful to the empiricists' core idea that all our knowledge about the world originates in sense experience.[1] According to Quine, however, the need for an analytic-synthetic distinction evaporates once we dismiss radical reductionism. After all, if "the unit of empirical significance is the whole of science," it becomes "folly to seek a boundary between synthetic statements, which hold contingently on experience, and analytic statements, which hold come what may" (TDE, 1951a, 42–43).[2]

[1] See (TDR, 1991b, 394): "I think Carnap's tenacity to analyticity was due largely to his philosophy of mathematics. One problem for him was the lack of empirical content: how could an empiricist accept mathematics as meaningful? Another problem was the necessity of mathematical truth. Analyticity was his answer to both."

[2] In "The Present State of Empiricism," written a few months after "Two Dogmas," Quine summarizes his twofold argument as follows: "[I]n 'Two dogmas of empiricism' I have argued that the various known ventures at explicating the notion of analyticity fail of their purpose, and I have suggested that the [. . .] distinction between the so-called analytic and the so-called synthetic [. . .] seems much less urgent and much less plausible when we withdraw to a holistic empiricism" (May 26, 1951, item 3015, my transcription). A transcription of "The Present State of Empiricism" is included in the appendix (A.5).

In chapter 5, I argued that "Two Dogmas" presents only a small step in the development of Quine's philosophy. If we examine the evolution of Quine's views between *Sign and Object* and *Word and Object*, I have argued, "Two Dogmas" was not the major breakthrough that it is often made out to be. Quine had been skeptical about the analytic-synthetic distinction for years, and he largely developed his positive contribution to epistemology in the years *after* "Two Dogmas."[3] As a result, Quine's most celebrated paper does not play a major role in the development of his naturalism; "Two Dogmas" should be primarily viewed as the paper in which Quine for the first time publicly advertises his still "tentative negative strictures on the analytic-synthetic distinction" (June 18, 1951, item 1200).

Still, we should not conclude that Quine's views about the analytic-synthetic distinction *themselves* are of little relevance to his metaphilosophical development. For, in blurring the boundary between matters of fact and matters of language, Quine indirectly challenged the supposed boundary between science and philosophy as well. After all, the strict distinction between science (conceived of as the study of fact) and philosophy (conceived of as the study meaning) that has dominated analytic philosophy for decades was grounded in the very distinction Quine became increasingly skeptical about in the 1940s.[4] Parallel to the events discussed in chapter 5, therefore, there is a second development that requires discussion if we are to provide a comprehensive reconstruction of the genesis of Quine's naturalism, viz. the evolution of his views on the analytic-synthetic distinction. In the present chapter, I will reconstruct this development.[5]

There has been a considerable debate about Quine's evolving views on analyticity in the literature. Richard Creath (1990, p. 31) argues that "[i]t was not until 1947, and then in private correspondence, that Quine came fully and finally to reject Carnap's doctrine that there are analytic truths," whereas Paolo Mancosu (2005, p. 331) points to a letter Quine wrote to J. H. Woodger in 1942, in which he argues that Carnap's "professedly fundamental cleavage between the

[3] What is more, Quine later realized that he should have arranged "Two Dogmas" differently: "my stress should have been on the second dogma, which is the real villain of the piece. So long as it is in sway, analyticity is needed to explain how mathematical and logical truths can be meaningful despite lack of empirical content. When on the other hand the second dogma is abandoned in favor of a holistic view [. . .] [a]nalyticity is no longer needed" (1987, item 3182, my transcription). Quine realized this fairly soon. Already in "The Present State of Empiricism" (see footnote 2), Quine writes that "the doctrine of reductionism [. . .] may be regarded as the more basic of the two [dogmas]" (May 26, 1951, item 3015, my transcription).

[4] Indeed, Burton Dreben, without doubt Quine's closest philosophical companion, basically equates the two distinctions. He has described Quine's mature view "as the mirror-image of Carnap's philosophy." According to Dreben, Quine "shows how Carnap is transformed once his most basic assumption is dropped, namely the fundamental distinction between philosophy and science, between the analytic and the synthetic" (1990, p. 88).

[5] Quine's evolving views on the science-philosophy distinction will be examined in chapter 7.

analytic and the synthetic is an empty phrase" (item 1237). Greg Frost-Arnold (2011, §5) defends an intermediate position. He argues that although Quine gave up on Carnap's *semantic* version of the analytic-synthetic distinction from the early 1940s onward, he "was not yet willing to commit himself to the radical view of 'Two Dogmas' until shortly before writing that piece." Isaac (2005, 2011) defends still another hypothesis. According to Isaac, Quine's motives for not publicly attacking the analytic-synthetic distinction until the early 1950s were largely political: "Up to the late 1940s, [Quine] had been content to mute his disquiet for the sake of presenting a united front on logical empiricism to the American academy" (2011, p. 274).

It is my contention, however, that these accounts lump together distinct aspects of Quine's development. Following his two-tiered argument in "Two Dogmas," I believe, we ought to distinguish between at least three questions about Quine's evolving views about the analytic-synthetic distinction:

(1) When did Quine start demanding a *behavioristically acceptable* definition of analyticity?
(2) When did he stop searching for such a definition of analyticity?
(3) When did he give up on the dogma of reductionism and conclude that we do not require an analytic-synthetic distinction in the first place?[6]

In the first part of the present chapter (section 6.2), I will answer these questions. For expository reasons, I will start with the last question (subsections 6.2.1–6.2.2), before examining the first two (subsections 6.2.3–6.2.6). Using new and previously unexamined evidence from the Quine archives, I argue that it is impossible to identify a specific moment in Quine's development at which he came to reject the analytic-synthetic distinction; each question requires an independent answer if we are provide a comprehensive account of his development.

In the second part of this chapter (section 6.3), I will reconstruct Quine's evolving views on the analytic-synthetic distinction *after* 1951. Looking back on the early 1950s, Quine claimed that he regretted his "needlessly strong statement of holism" in "Two Dogmas" (TDR, 1991b, 393). In the last sections, therefore, I examine the implications of Quine's claim that his early holism was needlessly strong (subsection 6.3.1) and challenge the common misconception that Quine changed his mind about scientific practice (subsection 6.3.2), the nature of logical truth (subsection 6.3.3), and the scope of his holism (subsection 6.3.4).

[6] Unfortunately, I have also conflated these questions myself at one point. See Verhaegh (2017a, §5).

6.2. Early Developments

6.2.1. Narrow and Wide Holism

Quine's second argument against the analytic-synthetic distinction, we have seen, relies on the dogma of reductionism. If we reject radical reductionism and accept Quine's holistic picture of theory testing, we do not *require* an analytic-synthetic distinction in the first place:

> [T]he second dogma creates a need for analyticity as a key notion of epistemology [. . .] the need lapses when we heed Duhem and set the second dogma aside. For given the second dogma, analyticity is needed to account for the meaningfulness of logical and mathematical truths, which are clearly devoid of empirical content. But when we drop the second dogma and see logic and mathematics rather as meshing with physics and other sciences for the joint implication of observable consequence, the question of limiting empirical content to some sentences at the expense of others no longer arises. (RGH, 1986b, 206–7)

If we accept Quine's "holistic empiricism" (May 26, 1951, item 3015, my transcription), in other words, we can account for logical and mathematical knowledge without positing an analytic-synthetic distinction.[7]

So let us look at the development of Quine's holism. When did Quine give up on radical reductionism and conclude that there is no need for an analytic-synthetic distinction in the first place? Reading Quine's work on the nature of scientific inquiry from the 1930s, one might get the impression that he was already committed to evidential holism in the earliest stages of his career. In his graduate school paper "Concepts and Working Hypotheses," for instance, Quine advances a view that seems pretty close to the holistic picture sketched in the last section of "Two Dogmas":

> If a recalcitrant item of experience, belonging to the field in question, should subsequently arise, modification somewhere in the system must take place, for it has been noted that a satisfactory conceptual system must accommodate every experience falling within the field. Thus it is that only the working hypothesis can stand which has endured without the emergence of any anomaly in the whole mass of experience since

[7] See also Quine's letter to Goodman: "In the case of analyticity, what I want to represent is *not merely* that the concept, if needed in epistemology, must be either posited outright or analyzed better than it has been to date; but that the very notion that a concept of analyticity is needed belongs to an epistemological orientation contrary to my own. I argue, in short, that to legitimize an analyticity concept is not even an agendum" (December 3, 1953, Goodman Archives, Harvard University Archives, accession 14359, unlabeled box, unlabeled folder, my transcription).

> its inauguration [. . .] one has a certain latitude as to where he may make his readjustments in the event of an experience recalcitrant to his system; and correspondingly there is some subjective option as to whether a chosen concept or a working hypothesis is to be branded as the point of 'error' in the antecedent system.[8] (March 10, 1931, item 3236)

Still, it would be a mistake to ascribe to the early Quine anything like the view he advanced in the 1950s, the crucial difference being that he did yet not apply his holism to our logical and mathematical knowledge.[9] Quine's holism was still of a narrow scope, applying only to the empirical sciences. Quine still aimed to explain the supposedly a priori character of logical and mathematical knowledge, like Carnap, in terms of analyticity.[10] Indeed, after first meeting Carnap in Prague in 1933, Quine writes: "I gained by the end of my five and a half weeks in Prague what I deemed to be a fair command of Carnap's philosophical and logistical theories [. . .] it has answered to my satisfaction the question of the epistemological status of mathematics and logic, a question formerly perplexing to me" (January 8, 1934, item 3254).

In "Truth by Convention," for instance, Quine still embraces an analyticity-based account of logical and mathematical knowledge:

> There are statements which we choose to surrender last, if at all, in the course of revamping our sciences in the face of new discoveries; and among these there are some which *we will not surrender at all*, so basic are they to our whole conceptual scheme. Among the latter are to be counted the so-called truths of logic and mathematics. [. . .] Now since these statements are destined to be maintained *independently of our*

[8] See also Isaac (2005, pp. 212–20) and Sinclair (2012, p. 342). Quine's way of thinking here can, to some extent, be traced back even to 1927, his first year in college (!): "Man uses as an outline for his knowledge the natural relationship of all things, so far as he has been able to determine that relationship in the incompleteness of his data. This web—to change the metaphor—which he has thus succeeded in partially spinning, he reinforces with synthetic thread of his own manufacture: to wit, the conventional classifications and man-made systems of compilation which form so large a part of human knowledge. These two kinds of relationship—the natural and the artificial—work together in such a way that often they are not to be distinguished one from the other" (March 10, 1927, item 3225).

[9] See also Hylton (2001, pp. 269–70) and Sinclair (2016, p. 87).

[10] Indeed, Carnap himself also combined his analyticity-based explanation of logical and mathematical knowledge with a narrow-scoped holism concerning the physical domain. See Carnap (1934, p. 318): "it is, in general, impossible to test even a single hypothetical sentence [. . .] *the test applies, at bottom, not to a single hypothesis but to the whole system of physics as a system of hypotheses.*" In a letter to Francisco Rodriguez-Consuegra, Quine argues that "holism was somewhat in the air back then" (February 19, 1992, item 923, my transcription).

> *observations of the world*, we may as well make use here of our technique of conventional truth assignment and thereby forestall awkward metaphysical questions as to our a priori insight into necessary truths. (TC, 1936, 102, my emphasis)

Although Quine here clearly does not want to invoke a metaphysical explanation of our supposedly a priori knowledge of logical and mathematical truths, he does not expand his holism to logic and mathematics so as to claim that our knowledge of those truths is ultimately a posteriori; he still believes that our logical and mathematical truths are maintained "independently of our observations of the world."

It is unclear when Quine for the first time considered expanding his holism to include logical and mathematical knowledge. Still, it is likely that Quine was confronted with this possibility during the Harvard Logic Group meetings of 1940–1941. For Alfred Tarski, one of the core members of the group, held a view that comes very close to what we might call a wide-scoped holism.[11] Already in 1930, a note in Carnap's diary shows, Tarski held that "there is only a [. . .] gradual and subjective distinction" between "tautological and empirical statements" (Haller, 1992). Furthermore, a letter Tarski sent to Morton White in 1944 shows that he also embraced the view that "no statement is immune to revision":

> [W]e reject certain hypotheses or scientific theories if we notice either their inner inconsistency, or their disagreement with experience, or rather with individual statements obtained as results of certain experiences. No such experience can logically compel us to reject the theory: too many additional hypotheses [. . .] are always involved. [. . .] Axioms of logic are of so general a nature that they are rarely affected by such experiences in special domains. However, I don't see here any difference 'of principle'; I can imagine that certain new experiences of a very fundamental nature may make us inclined to change just some axioms of logic. And certain new developments in quantum mechanics seem clearly to indicate this possibility. (White and Tarski, 1987, 31)

Looking back on his 1940–1941 discussions with Tarski and Carnap, Quine recalls how he and Tarski argued "persistently with Carnap over his appeal to analyticity" in the opening pages of *Introduction to Semantics* (TDR, 1991b, 392). Given the evidence available, I would suggest that it is reasonable to

[11] See also Mancosu (2005, §2).

conclude that Quine must have learned about Tarski's wide-scoped holism in this period as well.[12]

Despite Tarski's influence, however, the first evidence of Quine's wide-scoped holism is only from the spring semester of 1949, when he completed the first draft of *Methods of Logic* (then still called *Foundations of Logic*).[13] At some point in the writing process, Quine must have come up with his well-known solution to the problem of logical and mathematical knowledge. For *Methods of Logic* is the first document in which Quine embraces a wide-scoped holism:

> Physical objects are known to us only as parts of a systematic conceptual structure which, taken as a whole, impinges at its edges upon experience. [. . .] When [. . .] predictions of experience turn out wrong, the system has to be changed somehow. But we retain a wide latitude of choice as to what statements of the system to preserve and what ones to revise. [. . .] Our statements about external reality face the tribunal of sense experience not individually but as a corporate body.[14] (ML1, 1950a, xii)

[12] Indeed, in a 1945 letter to Skinner, Quine writes that "[t]he great Tarski [. . .] is one whom, unlike Carnap and Russell and Reichenbach, I consider genuinely sound and undeluded in his semantics and his philosophical orientation toward logic" (February 23, 1945, item 1001). See also Frost-Arnold (2011, p. 301): "It seems unlikely that Tarski never voiced these views about logic in Quine's presence during their year together at Harvard." Frost-Arnold argues that the Harvard Logic Group discussions were important for Quine's development in two different respects as well. First, Tarski presented to the group a proposal for a nominalist language, a language in which portions of arithmetic become synthetic, confronting Quine with the possibility of contracting the number of supposedly analytic truths. Second, the discussions revealed that Carnap had adopted a semantic approach to explicating analyticity, a move that conflicted with Quine's extensionalism, such that Quine came to "reject Carnap's then-current account of analyticity," which "perhaps made Quine even more suspicious in general of a notion he had begun to be skeptical about in 'Truth by Convention,' written when Carnap still accepted the extensional and syntactic approach" (p. 314).

[13] Still earlier versions of *Methods of Logic*, which were already widely used by students in mimeographed form, were titled *A Short Course on Logic* and *Theory of Deduction* (items 1401, 1467, 2829, and 2830). Quine started working on these logic texts after he postponed *Sign and Object*. See chapter 5. Indeed, Quine already seems to have been pondering the book project when he was still in the navy, as is evinced by a 1945 letter to Carnap: "such desultory work as I've done in leisure hours has gone toward planning one or another book, maybe yet another logic book and maybe a more philosophical venture, but with little tangible effect. My latest tendency is to think of the two ventures combined into one big book" (QCC, May 10, 1945, 378). It seems plausible that Quine decided to write just a logic book after he had shelved *Sign and Object*.

[14] The first glimmerings of such an alternative already appear in Quine's "Lectures on David Hume's Philosophy." Hume's philosophy inspires Quine to distinguish between two epistemological projects, one that reflects the "radical reductionism" he dismisses in "Two Dogmas," and one that comes remarkably close to the holistic position he adopts in *Methods of Logic* (here still called "pragmatism"):

> Constructive empiricism: explain all meaningful scientific discourse by contextual definition on the basis finally of reference to direct experience.

A few paragraphs later, Quine explicitly includes logical and mathematical knowledge:

> Mathematics and logic, central as they are to the conceptual scheme, tend to be accorded [. . .] immunity, in view of our conservative preference for revisions which disturb the system least; and herein, perhaps lies the 'necessity' which the laws of mathematics and logic are felt to enjoy. [. . .] But it must now be remarked that our conservative preference for those revisions which disturb the system least is opposed by a significant contrary force, a force for simplification. [. . .] Mathematical and logical laws themselves are not immune to revision if it is found that essential simplification of our whole conceptual scheme will ensue. (ML1, 1950a, xiii–xiv)

At some point in the spring semester of 1949, in other words, Quine adopted a wide-scoped holism. At most a few months later, Quine applied his newfound holism to the analytic-synthetic distinction. In response to an invitation by Goodman to come and talk about "Meaning" at the University of Pennsylvania, Quine writes: "Reasons against are that Mortie [Morton White] talked on the subject ('Analytic', which comes to the same thing) at the Fullerton Club last May, and I talked at Penn a couple of years ago, and you have no doubt been belaboring it this fall. [. . .] *But I expect there will be something new about my remarks this time*" (October 31, 1949, my emphasis).[15] Indeed, in his lecture, entitled "Animadversions on the Notion of Meaning," Quine gives an argument that come very close to his second argument in "Two Dogmas":

> Another view, not distinguishing the [linguistic and factual] components: we have our sense experience, and our own system of beliefs. [. . .] But it is underdetermined by experience. System as a whole must conform to experience along periphery; but disconformities can be repaired each by any of many changes of the system. We choose by two canons: 1) maximum elegance of whole system, 2) maximum conservationism. By 2), the more central principles resist change the more. These might be called the *more analytic*: matter of degree [. . .]

> Pragmatism: abandon such a project as impossible, and say that our discourse is merely variously conditioned by experience without being reducible to empirical terms. Abandon, therefore, empirical criticism of concepts; instead, judge any form of discourse in terms of its utility—this utility being measured within empirical science by ordinary empirical methods. (LDHP, 1946c, 135)

See Verhaegh (2017a, pp. 336–37). I thank James Levine for reminding me of this passage.

[15] Goodman Archives, Harvard University Archives, accession 14359, unlabeled box, unlabeled folder, my transcription.

> no *boundary* here between linguistic component and factual component.[16] (ANM, 1949, 155)

Still a few months later, in February 1950, Quine settled on his topic for his paper for the APA Eastern Division in Toronto. In a letter to Grace de Laguna, Quine writes: "Tentatively, my present notion is to attribute the sterility of current analytic philosophy to adherence to two myths: (1) the myth of a division [. . .] between truths of reason and truths of fact; (2) the myth that each statement of fact can, in isolation from the rest of science, be said to have its own peculiar phenomenal import." Instead, Quine wants to propose a "holistic empiricism, according to which it is not the individual statements of fact but only the collective statements of science as an interrelated totality that can significantly be brought into correlation with sensory evidence" (February 4, 1950, item 293).

6.2.2. Mathematical Objects and Mathematical Knowledge

So why did it take Quine until 1949 to conclude that we do not need an analytic-synthetic distinction to solve the problem of logical and mathematical knowledge? It is hard to provide a conclusive answer to this question; as far as I am aware, Quine simply did not directly write about the justification of logical and mathematical knowledge between 1941 and 1948.[17] Still, I do have a suggestion as to what *might* have blocked Quine's holistic solution until the late 1940s; viz. his nominalist project.

[16] It should be noted that there is still a small difference between "Animadversions" and "Two Dogmas." In "Two Dogmas," statements which we are not inclined to revise are not called "more analytic" than statements we easily give up. Because of this, the mature Quine would never claim that analyticity is a "matter of degree." But this is largely a matter of definition. In "Animadversions," Quine merely claims that we *might* call statements which we are less inclined to revise "more analytic." If Quine had adopted this policy in "Two Dogmas," he would also have concluded that the analytic-synthetic distinction is a matter of degree. See Quine's letter to James Cowan: "My position on analyticity was flat rejection of the distinction, rather than gradualism. Reluctance to give up a sentence, in the course of revising a theory, is indeed a matter of degree: but I do not identify analyticity with this" (March 12, 1975, item 234).

[17] One hypothesis is that Quine simply was not concerned with the problem. Indeed, Burton Dreben has pointed out that although Harvard logicians where generally preoccupied with the "nature of logical justification" in the 1920s and 1930s, "[n]o questions of justification arise" in Quine's work on logic in the early 1930s (1990, 83). I am sympathetic to Dreben's thesis, but I do not believe that it extends to Quine's work in the 1940s. After all, Quine lists the problem as one of most urgent questions in philosophy in his letter to Rockefeller Foundation (see section 5.7). In this letter, Quine writes: "Clarification of the notion of cognitive meaning, or of the relation of cognitive synonymy of phrases, is needed in order to make sense of the distinction between analytic and synthetic judgments. This distinction has played a central rôle in philosophy from Kant to Carnap, and is of vital importance to the philosophy of mathematics" (October 20, 1946, item 921).

As we have seen in section 5.2, Quine intensified his attempts to "set up a nominalistic language in which all of natural science can be expressed" (DE, 1939b, 708) in the years immediately following the Harvard Logic Group meetings—attempts that eventually culminated in his joint paper "Steps toward a Constructive Nominalism" (Goodman and Quine, 1947). We can get a pretty good overview of Quine's ideas in this period from his 1946 lecture on nominalism. In the lecture, Quine distinguishes between a "mental" and a "physical" version of nominalism, the former allowing only mental and the latter allowing only physical particulars:

> In the mental case [the nominalist's] motive may be an extreme sensationalism: what we are presented with are sensory events, and it is unphilosophical to assume entities beyond them, *in particular universals.* In the physical case, his mentality is likely to be that of Lord Kelvin, who insisted that he did not understand a process until it was reduced to terms of impact of bodies like billiard balls. [. . .] Modern physics may seem to have cut the ground from under this physical type of nominalist, in abandoning even Kelvin's billiard balls [. . .] [b]ut the nominalist is capable of surviving this [. . .] [T]he nominalist reserves the right to refurbish this conceptual scheme [. . .] and to produce a substitute conceptual scheme which, while still theoretically adequate to the physicist's purposes, will not countenance any entities beyond those whose existence it is within the *physicist's* professional competence to assert. (NO, 1946a, 17–18)

For our present purposes, it is important to note that especially physical nominalism—the variant adopted in Goodman and Quine's joint paper—presupposes a distinction between physical and mathematical objects. Where the holistic empiricist holds that *both* physical and mathematical objects are posits (though unimpeachably real), which serve to simplify our conceptual scheme, the nominalist presupposes that mathematical objects are somehow more problematic than physical objects. After all, the nominalist aims to reduce mathematical objects to something less problematic without being equally concerned about the status of physical objects.[18]

[18] It should be noted that there remains a small difference between the two types of objects in Quine's mature view; he would welcome a reduction of abstract objects to concrete objects for simplicity's sake, but he does not accept a reduction of concrete objects to universals. See (WBP, 1981c). This is not an epistemological difference, however. In a 1958 letter to Church, he jokingly compares his mature attitude toward mathematical objects with mysogynism: "I do have an attitude toward universals somewhat like that of misogynists towards women; but also I tend to go along with actual misogynists in acknowledging that the deplored objects are there" (May 14, 1958, item 224).

In March 1948, however, Quine changes his mind. Perhaps dissatisfied with the results of his nominalistic endeavors,[19] Quine adopts a new approach to abstract objects. In a letter to J. H. Woodger, Quine writes about his "rapidly evolving" views:

> I suppose the question what ontology to accept is in principle similar to the question what system of physics or biology to accept: it turns finally on the relative elegance and simplicity with which the theory serves to group and correlate our sense data. [. . .] Now the positing of abstract entities (as values of variables) is the same kind of thing. As an adjunct to natural science, classical mathematics is probably unnecessary; still it is simpler and more convenient than any fragmentary substitute that could be given meaning in nominalistic terms. Hence the motive—and a good one—for positing abstract entities (which classical mathematics) needs. [. . .] These very relativistic and tolerant remarks differ in tone from passages in my paper with Goodman and even in my last letter, I expect. My ontological attitude seems to be evolving rather rapidly at the moment.[20] (March 22, 1948, item 1237)

In March 1948, in other words, Quine starts arguing that physical and mathematical objects are epistemically on a par. Two months later, he submits "On What There Is" to the *Review of Metaphysics* (May 11, 1948, item 1443), claiming that "[t]he analogy between the myth of mathematics and the myth of physics is [. . .] strikingly close" (OWTI, 1948, 18). And although Quine still writes Hugh Miller that he does not know "what it would mean to say [. . .] that arithmetic is *not* analytic" in May 1948 (item 724, my transcription),[21] a few months later Quine writes his introduction to *Methods of Logic,* arguing that mathematical and logical laws "are not immune to revision" either (ML1, 1950a, xiv).

Although it is difficult to offer any conclusive evidence for the hypothesis, in other words, it is my contention that Quine's adoption of a holistic empiricism in the late 1940s was made possible by his changing attitude on nominalism. After all, if physical and mathematical objects have the same epistemic status, then it cannot be a surprise that we do not need a special account (i.e. an analytic-synthetic distinction) to account for logical and mathematical knowledge either.

[19] In his autobiography, Quine explains how he and Goodman failed to give a complete nominalist account of proof theory, which assumes "strings of [s]igns without limit of length, whereas our program could countenance them only insofar as physically realized" (TML, 1985a, 198).

[20] See also Mancosu (2008, p. 43).

[21] See section 5.7.

6.2.3. The Principle of Tolerance

If we adopt Quine's evidential holism, we have seen, we do not *require* an analytic-synthetic distinction in the first place. Quinean holism implies that we can give a unified account of scientific inquiry and that we do not need a special explanation of logical and mathematical knowledge. I have argued that Quine, perhaps influenced by Tarski and his "rapidly evolving" views on nominalism, starts rejecting radical reductionism in 1949.

In contemporary discussions of the Carnap-Quine debate, scholars often focus on this second argument against the analytic-synthetic distinction. After all, it seems that if the second argument is correct, we can solve the Carnap-Quine debate without quibbling about the analytic-synthetic distinction itself. Historically, this assessment seems to be correct. Much of the appeal of Quine's case against the analytic-synthetic distinction seems to derive from his argument that no such distinction is required to account for logical and mathematical knowledge. Indeed, despite his focus on the first argument in "Two Dogmas," we have seen that Quine himself also viewed the dogma of radical reductionism as "the real villain of the piece" in later stages of his career.[22]

Dialectically, however, Quine's second argument does not suffice to decide the Carnap-Quine debate. It merely shows that, epistemologically, there is no *need* for an analytic-synthetic distinction, not that there *is* no such distinction. Carnap could easily grant Quine his holistic picture of inquiry and still maintain an analytic-synthetic distinction; he could simply acknowledge that although no statement is immune to revision, there are two *kinds of revisions*: changes of theory and changes of meaning. In fact, this is precisely what he argued in "Quine on Analyticity"—a reply Carnap wrote in 1954 but that remained unpublished until the early 1990s:

> Quine has emphasized that in revising the total system of science no statement and no rule is immune or sacrosanct. [. . .] Thus far I agree with Quine. [. . .] However, I cannot agree with Quine when he concludes about this that there is no sharp boundary between physics and logic. [. . .] Since the truth of an analytic sentence depends on the meaning, and is determined by the language rules and not the observed facts, then an analytic sentence is indeed 'unrevisable' in another sense: it remains true and analytic as long as the language rules are not changed. (1954, pp. 432–33)

Carnap, in other words, argues that an analytic-synthetic distinction is perfectly compatible with Quine's holistic empiricism. If one solely focuses on Quine's

[22] See footnote 3.

second argument, one ends up with two competing pictures of inquiry: one which blurs the boundary between changes of fact and changes of meaning and one which maintains the distinction.

In view of Carnap's response, one might conclude that the difference between Carnap and Quine is merely pragmatic—that it is up to all of us to decide which picture of inquiry is most appealing. The Quinean will argue that his picture is simpler because it presents a unified account of inquiry, whereas the Carnapian will maintain that his picture does more justice to the working scientist's intuitions about logic and mathematics.[23] In view of Carnap's principle of tolerance, one might even conclude that we can set up languages in both ways. For some purposes an analytic-synthetic distinction might be useful; for other purposes we do not need one.[24]

To conclude that the difference between Carnap and Quine is merely pragmatic, however, is to ignore Quine's first argument—it is to ignore his claim that the analytic-synthetic distinction lacks a behavioristically adequate definition. Quine, as we have seen in chapter 4, poses a severe constraint on artificial languages; if the analytic-synthetic distinction is not grounded in natural language, it also cannot be created in an artificial one: "Semantical rules determining the analytic statements of an artificial language are of interest only in so far as we already understand the notion of analyticity; they are of no help in gaining this understanding" (TDE, 1951a, 36). Unlike Quine's second argument against the analytic-synthetic distinction, in other words, his demand for an empirical clarification of notions like analyticity *does* reveal a substantive difference between

[23] Indeed, in response to the supposed intuitive appeal of the analytic-synthetic distinction, Quine often attempted to explain away those intuitions. See, for example, (PT, 1990a, 15): "If asked why he spares mathematics, the scientist will perhaps say that its laws are necessarily true, but I think we have here an explanation, rather, of mathematical necessity itself. It resides in our unstated policy of shielding mathematics by exercising our freedom to reject other beliefs instead." Similarly, in a letter to Paul Horwich, Quine attempts to explain why we tend to attribute necessity to logical truths but not to common sense truths like "There have been black dogs," which we also shield in the light of adverse experience: "I think the reason is that these [common sense certainties] are near to observation, so that people are less mystified about their justif'n. In the case of math'l truth they invoke the mystical notion of metaph'l necessity. They haven't thought of the common-sense expl'n that holism offers" (January 2, 1992, item 530, my transcription).

[24] Carnap, himself, of course would argue that we need the distinction for *scientific* purposes. See Carnap (1966, pp. 257–58): "The theory of relativity [. . .] could not have been developed if Einstein had not realized that the structure of physical space and time cannot be determined without physical tests. He saw clearly the sharp dividing line that must always be kept in mind between pure mathematics, with its many types of logically consistent geometries and physics, in which only experiment and observation can determine which geometries can be applied most usefully to the physical world." Quine, on the other hand, uses Einstein to favor his own account in "The Present State of Empiricism" (item 3015). See appendix section A.5.

Carnap and Quine—i.e. a fundamental disagreement about the nature of language. Carnap adopts the principle of tolerance; Quine rejects it.[25]

6.2.4. A Behavioristically Acceptable Definition

So let us return to Quine's development. When did Quine start demanding a behavioristically acceptable definition of analyticity? When did he develop what I have called a strictly behaviorist variant of empiricism?[26] To a certain extent, it is not very difficult to determine the roots of Quine's behavioristic empiricism. From the very beginning of his career, Quine was a determined empiricist; nowhere does he question its plausibility or even take seriously alternative positions. In fact, on the few occasions where he looks back on his intellectual development, Quine suggests that he was committed to a strictly behaviorist variant of empiricism from the very start. Reflecting on his dismissal of intensional notions in Whitehead and Russell's *Principia Mathematica* during his final year at Oberlin College, for example, Quine notes:

> The distrust of mentalistic semantics that found expression in "Two Dogmas" is thus detectable as far back as my senior year in college. Even earlier I had taken kindly to John B. Watson's *Psychology from the Standpoint of a Behaviorist*, which Raymond Stetson had assigned to us in his psychology class. Nor do I recall that it shocked any preconceptions. It chimed in with my predilections.[27] (TDR, 1991b, 390)

Quine's early commitment to a strictly behaviorist variant of empiricism is also apparent in his approach to analyticity in the 1930s. We have seen that Quine accepted an analytic-synthetic distinction in the mid-1930s. Still, he was already implicitly explicating the notion in behavioral terms. Even in his 1934 "Lectures on Carnap," for instance—lectures he would later describe as "completely uncritical" (EBDQ, 1994c, 153)[28]—Quine proposes that we render

> only such sentences analytic as we shall be most reluctant to revise when the demand arises for revision in one quarter or another. These

[25] In a letter to Quine, Daniel Isaacson recalls that Quine has once described the principle of tolerance as "the serpent in Carnap's garden" in a response to his paper at a conference in 1988 (July 21, 1992, item 553).

[26] See section 4.2.

[27] See also Quine (TML, 1985a, 59) and (IA, 1986a, 7). For a discussion of the way in which Quine's student papers already advertise a distinctively empiricist picture of inquiry, see Isaac (2005) and Sinclair (2012). For a discussion of Quine's commitment to behaviorism as a student, see Verhaegh (forthcoming-d).

[28] See also chapter 5, footnote 10.

> include all the truths of logic and mathematics; we plan to stick to these in any case, and to make any revisions elsewhere.[29] (LC, 1934b, 63)

Whereas analyticity always served an epistemic function for Carnap (according to the mature Quine)—explaining why our logical and mathematical statements are *justified*—Quine here interprets the concept in strictly *psychological* terms; we call the truths of logic and mathematics analytic because it is a psychological fact that we will not give them up in the light of adverse experience.[30]

Still, when we look at Quine's work from the 1930s, it is remarkable that many of his writings about the analytic-synthetic distinction only *implicitly* rely on this behavioristic picture of language. Even though Quine interprets the distinction behavioristically, in other words, he does not yet seem to realize that this is a different approach than Carnap's.[31]

In the 1940–1941 Harvard Logic Group discussions, as we have seen, Quine adopts a more critical approach to Carnap's analytic-synthetic distinction.[32] Still, even in these Logic Group discussions, Quine does not seem to be fully aware of the nature of his disagreement with Carnap. As Frost-Arnold notes in his discussion of Carnap's summaries of these meetings, Quine did not yet seem to have a completely worked out argument against the analytic-synthetic distinction.

[29] See also Carnap's notes of his discussion with Quine in Prague 1933, first discovered by Neil Tennant (1994). In the note, Carnap reports that Quine responds to his distinction between "logical axioms and empirical sentences" by claiming that logical statements "are the sentences we *want* to hold fast" (translation by Quine, TDR, 1991b, 391, my emphasis).

[30] See also, for example, Creath (1987, pp. 485–86) and Hylton (2001). I follow Creath in describing Quine's characterization as "psychological" because, technically, Quine does not mention behaviorism in his lectures. Still, this is what underlies Quine's characterization. In "Truth by Convention," a paper that largely resembles his first lecture on Carnap, for example, Quine explicitly writes that the apparent contrast between a priori and a posteriori truths (and thus the analytic and the synthetic) retains reality "*behavioristically* [. . .] as a contrast between more and less firmly accepted sentences" (TC, 1936, 102, my emphasis). Interestingly, Frost-Arnold (2011, p. 300 n. 15) suggests that Quine's identification of the a priori with claims that can be held true come what may might be influenced by C. I. Lewis, who held that the a priori is that "which we can maintain in the face of all experience, come what will" (1929, p. 231). Recently, Lewis's influence on Quine's development has received quite some attention in the literature. See, for example, Baldwin (2007, 2013), Sinclair (2012, 2016), Murphey (2012, ch. 1), and Morris (forthcoming).

[31] In fact, it seems that Quine still thinks that Carnap would agree with him as he explicitly notes that his definition of the a priori explicates the notion "without reference to a metaphysical system" (TC, 1936, 102). See also Ebbs (2011a) and Morris (forthcoming), who argue that "Truth by Convention" in general should not be read as an argument against Carnap views.

[32] See (TML, 1985a, 149–50): "By way of providing structure for our discussions, Carnap proposed reading the manuscript of his *Introduction to Semantics* for criticism. Midway in the first page, Tarski and I took issue with Carnap on analyticity. The controversy continued through subsequent sessions, without resolution and without progress in the reading of Carnap's manuscript."

> Quine, unlike Tarski, does not articulate complete arguments against Carnapian analyticity in the notes; rather, he simply voices disagreement with two of Carnap's core commitments. [. . .] The first difference between the two is that Carnap holds sentences of the form '*p* is analytic' to be themselves analytic, whereas Quine considers them synthetic. Second, Quine considers Carnap's characterization of analyticity in modal terms fundamentally unclear. (2013, p. 89)

This second difference, Frost-Arnold argues, is mostly an argument against Carnap's *intensional* treatment of analyticity, not an argument against analyticity in general. Even though Quine argues that Carnap's latest characterization of analyticity is "fundamentally unclear," in other words, he does not yet seem to realize what grounds his disagreement with Carnap.[33]

It is my contention that this changes two years later, when Carnap and Quine resume their correspondence after eighteen months of silence. In his first letter to Carnap after the Logic Group meetings, Quine writes:

> Thank you very much for [*Introduction to Semantics*]. I'm impressed with it as a masterly job of organization and presentation, and much of the theory is decidedly to my liking despite my dissension on certain points. On the other hand I do feel that the points where I dissent are peculiarly crucial to semantics; and my mind has become somewhat clearer on them in the year and a half since we talked. (QCC, January 5, 1943, 294)

In his letter, Quine explains that he has found a way to define analyticity in terms of synonymy,[34] but maintains that synonymity itself requires clarification as well:

> The problem remains, of course, to explain this basic synonymity relation. This is a relation whose full specification, like that of designation, would be the business of pragmatics. [. . .] The definition of this relation of synonymity, within pragmatics, would make reference to criteria of behavioristic psychology and empirical linguistics. [. . .] I find it interesting to have reduced the notion of analytic to [. . .] synonymity because I feel this shows, more clearly than hitherto, the gap that has to be bridged. (QCC, January 5, 1943, 298)

[33] See also Ebbs (2016, p. 34), who writes that Quine's motivation for rejecting Carnap's approach "was probably not fully clear to him in 1939–1941."

[34] See Quine's "Notes on Existence and Necessity," which appeared later that year: "we can define an analytic statement as any statement which, by putting synonyms for synonyms, is convertible into an instance of a logical form all of whose instances are true" (NEN, 1943b, 120).

Quine, in other words, has become clearer about what needs to be done if we are to have an acceptable analytic-synthetic distinction. What Quine does not yet seem to realize, however, is that Carnap does not share his call for a behavioristic definition. Indeed, Carnap replies:

> Here is an important methodological point. I believe that we cannot construct an exact and workable theory of concepts like 'true', 'analytic', 'meaning', 'synonymous', 'compatible' etc. if we refer merely to the actually used language of science. It seems to me that we can use those concepts only if we replace the given language by a system of rules; in other words, we have to go from pragmatics and descriptive semantics to pure semantics. [. . .] If the concept 'synonymous' is to be used at all in pure semantics, you have to state rules for it.[35] (QCC, January 21, 1943, 309)

This exchange pretty well sums up the above-discussed distinction between Carnap and Quine. Whereas Carnap believes that we can simply define a concept by stating rules for its use, Quine argues that a concept is useless if it is not grounded in natural language. Whereas Carnap thinks of artificial languages as mathematical structures, for Quine artificial languages are only intelligible as hypothetical natural languages:[36]

> Artificial languages are in no way different from natural ones [. . .] except that in the case of artificial languages we have a hypothetical construction *as if* there were people speaking it; the question of a behavioristic definition of 'analytic' [. . .] remains basic—part of the pragmatic substructure of the (more abstract) science of semantics.[37] (QCC, May 19, 1943, 338)

[35] Later that year, Church invokes a related methodological objection to Quine's suggestion: "as I understand you, your proposal is based not on an abstract notion of language but on an empirical one [. . .] I think there are [. . .] serious objections [. . .] of vagueness in the notion of an empirical language which defeat its use for defining" (November 8, 1943, item 224).

[36] See also Frost-Arnold (2013, pp. 62–63): "Quine holds adamantly to the assumption that our language is part of same material world we inhabit and our sciences investigate. [. . .] Carnap, on the other hand, studies language primarily as a mathematical subject instead of a natural one."

[37] In discussing a similar issue, Frost-Arnold (2013, p. 105) concludes that from a Carnapian perspective, Quine's position is analogous to asking, "Which is correct: pure geometry or applied geometry?" and suggests that Quine would respond by saying that "[a] given mathematical system wouldn't deserve the name 'geometry' at all, if it did not admit of an interpretation involving spatio-temporal magnitudes in the empirical world." Coincidentally, Quine puts the issue in precisely these terms in a letter to Church: "In geometry [. . .] we understand 'line' by extension or idealization or schematization of the properties of tightness an thinness in a string. [. . .] Now the case is parallel for language. The artificial formal languages [. . .] are idealizations or schematizations of language in the more direct empirical sense studied by field linguists" (November 13, 1943, item 224).

Of course, there is nothing new in this analysis. It will be familiar to anyone who has closely studied "Two Dogmas" (especially section 4). It is my contention, however, that this *was* new to Quine in 1943. In their debate about the necessity of a behavioristically acceptable definition of synonymy, Quine for the first time realized *why* he and Carnap disagreed. In a letter to Alonzo Church, Quine explains:

> [M]y attitude toward 'formal' languages is very different from Carnap's. Serious artificial notations, e.g. in mathematics or in your logic or mine, I consider supplementary but integral parts of natural language. [. . .] Thus it is that I would consider an empirical criterion [. . .] a solution of the problem of synonymy in general. And thus it is also that [. . .] I am unmoved by constructions by Carnap in terms of so-called 'semantical rules of a language'.[38] (August 14, 1943, item 224)

6.2.5. Triangular Correspondence

Let me sum up what we have established thus far. If we are to reconstruct the development of Quine's view on analyticity, we have to answer three questions: (1) When did Quine start demanding a *behavioristically acceptable* definition of analyticity? (2) When did he stop searching for a such definition of analyticity (3) When did he give up on radical reductionism, concluding that there is no need for an analytic-synthetic distinction in the first place? I have answered the first and the last question. I have argued that although Quine always defended a narrow-scoped holism and a strictly behaviorist variant of empiricism, he became aware of the nature of his disagreement with Carnap only in 1943, realizing that he, unlike Carnap, demanded a behavioristically acceptable definition of analyticity. Furthermore, I have argued that Quine adopted his wide-scoped holism in 1949, plausibly due to the influence of Tarski and his "rapidly evolving" views about the epistemic status of mathematical objects.

This leaves us with the second question: when did Quine stop searching for a behavioristically acceptable definition of analyticity? In his introduction to the Carnap-Quine correspondence, Richard Creath has argued that "[i]t was not until 1947" that Quine "came firmly and finally to reject the notion of analyticity as unintelligible" (1990, p. 31), a conclusion he justifies using Quine's "triangular correspondence with Nelson Goodman and Morton White" in the summer of 1947 (p. 35). Isaac (2011, p. 275) and Murphey (2012, p. 77) also emphasize the correspondence's influence on Quine's development. So before I continue my discussion, let me briefly discuss this exchange of letters.

[38] See also section 4.7.

In May 1947, White asked Quine to comment on a manuscript that he would later publish as "On the Church-Frege Solution of the Paradox of Analysis" (1948). Briefly put, the paradox runs as follows. Consider the following two statements:

(i) The attribute of being a brother is identical with the attribute of being a male sibling.
(ii) The attribute of being a brother is identical with the attribute of being a brother.

Intuitively, (i) is informative whereas (ii) is not. Yet, if (i) is true, then both statements say the same thing.[39]

Quine, who had already corresponded with White on the paradox in 1945 (item 1213), suggests that the puzzle might be solved using C. I. Lewis's distinction between intensional and structural synonymy.[40] The problem with this solution, however, is that Quine did not know of any behavioristically acceptable definition of intensional synonymy. Still, Quine's letter shows that he had not yet given up hope of finding such a definition in 1947:

> It's bad that we have no criterion of intensional synonymy; still, this frankly and visibly defective basis of discussion offers far more hope of clarity and progress, far less danger of mediaeval futility, than does the appeal to attributes, propositions, and meanings.[41] (GQW, June 3, 1947, 339–40)

Goodman, however, defended a much more stringent position than Quine. In response to both White and Quine, Goodman argues that "the lack of any behavioristic criterion (or even the dimmest suggestion as to how one might be set up) is a sign that we are not at all clear as to what it is that we have to define" (GQW, June 8, 1947, 343). According to Goodman, the whole project of seeking

[39] In one letter, Quine shows that the paradox can also be formulated without invoking attributes: "An 'analysis' has the form $\zeta = \eta$, where ζ and η are synonymous; therefore the whole analysis is synonymous with, or translatable into, the triviality $\zeta = \zeta$" (GQW, June 3 1947, 339).

[40] Two statements are intensionally synonymous when they have the same intension, whereas structural synonymy is a narrower relation which depends on the statements' constituents and their syntactic order. All analytic statements have the same (null-)intension, but not all analytic statements are structurally synonymous. See Lewis (1946) and Carnap (1947). Appealing to this distinction solves the paradox according to Quine because (i) and (ii) above are intensionally but not structurally synonymous.

[41] See also the 1946 lecture "On the Notion of an Analytic Statement," where Quine argues that although there are real problems with the notion of analyticity, his "attitude is not one of defeatism" (ONAS, 1946b, 35).

acceptable definitions of analyticity and synonymy has to be abandoned: "when Van [Quine] uses a term and hopes for a behavioristic criterion he can't vaguely outline, he is employing a meaningless mark or noise on the ground that he needs it [. . .] and hopes that a meaning will be found for it" (p. 343).

Quine, obviously not very happy to be placed in the intensionalist camp by Goodman ("I have always been all for extension, with the world against me"), responded by backing Goodman's position. Quine now granted that he also "doesn't know how to apply 'analytic', much less define it" (GQW, July 6, 1947, 353–54). Goodman then, in his final letter, urges Quine to give up on the project of defining analyticity:

> If Van agrees that he not only doesn't know how to define "analytic" but doesn't know how to apply it either, what is it that he is hoping to find a behavioristic definition for? [. . .] he is looking for a behavioristic definition for which the test of adequacy will presumably be in accordance with a usage which he doesn't have before him. (GQW, July 8, 1947, 356–57)

6.2.6. Quine on Asemiotism

Richard Creath suggests that Quine abandoned his search for a behavioristically acceptable definition of analyticity in the wake of his triangular correspondence:

> [I]n 1943 Quine had not yet given up on analyticity. Plainly, he did thereafter. The most likely date that he did so would be the summer of 1947. At that time Quine engaged in a triangular correspondence with Nelson Goodman and Morton White. Even then Quine was cautious, almost reluctant. But over the course of that summer he clearly came firmly and finally to reject the notion of analyticity as unintelligible and to eschew it in his own discussions of language. One result of those 1947 discussions was Morton White's 1950 paper "The Analytic and Synthetic: An Untenable Dualism." Quine could hardly let others speak for him, however, so when he was asked to give an address to the American Philosophical Association, the product was "Two Dogmas of Empiricism." (1990, p. 35)

Despite Creath's suggestion, however, there is no evidence that Quine actually abandoned his attempts to find a behavioristically acceptable definition of analyticity in the triangular correspondence. The exchange between Goodman, White, and Quine ends with Goodman's remark that Quine is in "the same position that he would be if he were to set out to define the Calubrian word 'phwanischk' " and that the only difference between them "is perhaps over the illusory aid and comfort he seems to give the enemy" (GQW, July 8, 1947, 356–57). Indeed, Quine himself in his last letter argues that "the status of my doctrine of intensional synonymy remains unchanged" (GQW, July 6, 1947, 354), referring back to his earlier claim that

> isolated, the problems [of meaning, analysis, analytic truth, necessity, etc.] take the form essentially of a single problem: behavioristic definition of intensional synonymy. I am not saying this problem can be solved; what I say is that this is the way the problems under consideration should be conceived, if at all. And that I can't understand such things as the philosopher's (e.g. Carnap's) concept of 'analytic truth' unless this problem is solved.[42] (GQW, June 27, 1947, 345)

If we solely focus on the triangular correspondence, in other words, we cannot draw the conclusion that Quine definitively abandoned attempts to find a behavioristically acceptable notion of analyticity during the summer of 1947.

Fortunately, the Goodman archives *do* contain evidence about Quine's views after 1947. And perhaps surprisingly, this evidence reveals that Quine did not give in to Goodman's pressure at all. For a previously unpublished letter from Quine to Goodman shows that Quine had not completely given up on defining analyticity even in 1949, the year before he wrote "Two Dogmas." In the letter, Quine responds to Goodman's suggestion that "[t]wo terms have the same meaning if they have the same extension" (January 16, 1949).[43] In his reply, Quine dismisses Goodman's suggestion and reminds him of his position in 1947 (a position he dubs "asemiotism"):

> There is another doctrine, not to be confused with extensionalism (whereof we are both adherents), which might be named *asemiotism.* It is the doctrine that no coherent notions of meaning, analyticity, etc. are to be hoped for. While *I do not identify myself with asemiotism,* I must concede it a certain plausibility. I used to think of you, indeed, as one of its staunchest proponents; and I urge you to reconsider your recent abandonment of the doctrine. For you do evidently abandon it; far from repudiating the concept of meaning as a good asemiotist should, you now espouse a theory of meaning according to which the meaning of a name is the thing it names and the meaning of a general term is its

[42] In his last letter Quine refers back to this paragraph as "the long paragraph of the first page of my letter of June 27" (GQW, July 6, 1947, 354). In White's *A Philosopher's Story*, however, the letter is dated June 30. Still, it is clear that Quine is referring to this paragraph. Quine's private copy of this letter *is* dated June 27 (item 1213). Presumably, White received the letter on June 30.

[43] Papers of Nelson Goodman, Harvard University Archives, accession 14359, unlabeled box, unlabeled folder.

> extension. This, far from being asemiotism, is merely a brand of neo-Confusionism.[44] (January 27, 1949, item 420, my emphasis)

Quine's remark that he himself does not accept asemiotism shows that he had not completely given up on the project of finding a behavioristically acceptable definition of analyticity as late as 1949.

So when *did* Quine give up on this project? Perhaps surprisingly, it is my contention that the answer should be that he never did. To see this, reconsider Quine's summaries of his argument against the analytic-synthetic distinction in the early 1950s:

> I have argued that the various known ventures at explicating the notion of analyticity fail of their purpose, and I have suggested that the [. . .] distinction between the so-called analytic and the so-called synthetic [. . .] seems much less urgent and much less plausible when we withdraw to a holistic empiricism. (May 26, 1951, item 3015, my transcription)

> [W]hat is ["Two Dogmas"]? (a) The observation that the analytic-synthetic distinction has never been adequately def'ned, though all too widely taken for granted. (b) The tentative conjecture that epistemology might develop more fruitfully under some very different sort of conceptualization. (June 18, 1951, item 1200, my transcription)

> In the case of analyticity, what I want to represent is *not merely* that the concept, if needed in epistemology, must be either posited outright or analyzed better than it has been to date; but that the very notion that a concept of analyticity is needed belongs to an epistemological orientation contrary to my own.[45] (December 3, 1953).

In each case, Quine does not say that there is *no* behavioristically acceptable definition of analyticity, he merely claims that *existing* accounts fail and that the notion should be analyzed *better*. Although Quine started to recognize that there is no *need* for an analytic-synthetic distinction after 1948 (see section 6.2), he simply never dismissed the possibility of finding a behavioristically acceptable definition of analyticity. After all, his arguments from the 1950s are still compatible with his earlier position that "I can't understand such things as the philosopher's (e.g. Carnap's) concept of 'analytic truth' unless this problem [of finding a behavioristic definition] is solved."

[44] Both Goodman's letter and Quine's response are included in the appendix (A.4).

[45] Goodman Archives, Harvard University Archives, accession 14359, unlabeled box, unlabeled folder.

It is not my aim here to suggest that Quine did not reject the analytic-synthetic distinction. He certainly did, not in the least because he was convinced that no such distinction was *needed*. In "Two Dogmas of Empiricism," after all, Quine dubs the analytic-synthetic distinction "an unempirical dogma of empiricists, a metaphysical article of faith" (TDE, 1951a, 37). His challenge to Carnap, however, was not merely to find a behavioristically acceptable explication of analyticity. Indeed, Quine found such explications himself in *Word and Object* and *The Roots of Reference*:

> I call a sentence *stimulus-analytic* for a subject if he would assent to it, or nothing, after every stimulation (within the modulus). (WO, 1960, 55)
>
> [I]f every member of a language community learns that a certain sentence is true by learning how to use one or more of its component terms, then there are no obstacles toward counting the sentence as true in virtue of meaning. (RR, 1973, §21)

Rather, his challenge to Carnap was to find a definition of analyticity that is both behavioristically acceptable *and* can play the role Carnap believed the analytic-synthetic distinction should play.[46]

Quine, in sum, never needed to give up on the attempt to find a behavioristically acceptable definition of analyticity. Although Quine's *demand* for such a definition reveals a substantial difference between him and Carnap (section 6.4), his argument *against* Carnap's epistemological use of the distinction did not require him to proclaim that no such explication of analyticity can be found.

6.3. After "Two Dogmas"

Now that I have mentioned Quine's definitions of analyticity in *Word and Object* and *The Roots of Reference*, we enter a new a phase in our discussion of his development. For despite the tremendous influence of his arguments in "Two Dogmas," Quine seems to have changed his mind about several components of his holistic empiricism in later stages of his career. Having dismissed one such apparent change of mind—i.e. the misconception that Quine's "vegetarian"

[46] In (WO, 1960, 67), Quine calls "stimulus analyticity" his "strictly vegetarian imitation" of the "sweeping epistemological dichotomy between analytic truths as by-products of language and synthetic truths as reports on the word." Similarly, in (RR, 1973, 80) Quine argues that "we have here no such radical cleavage between analytic and synthetic sentences as was called for by Carnap and other epistemologists." See Hylton (2002). Note that the definitions in *Word and Object* and the *Roots of Reference* are not the same. The first, unlike the latter, implies that all true standing sentences are stimulus-analytic. Hence the first definition works only for observation sentences.

notions of analyticity undermine his conclusions in "Two Dogmas"—I will use the second part of this chapter to dismiss a more widespread misunderstanding about his later development, viz. the misconception that Quine changed his mind about the *scope* of his holistic empiricism.

Although wide-scoped holism plays an important role in his argument against the analytic-synthetic distinction, as we have seen (section 6.2), Quine explicitly claims that he regrets his "needlessly strong statement of holism" in "Two Dogmas":

> Looking back on it, one thing I regret is my needlessly strong statement of holism. [. . .] In later writings I have invoked not the whole of science but chunks of it, clusters of sentences just inclusive enough to have critical semantic mass. By this I mean a cluster sufficient to imply an observable effect.[47] (TDR, 1991b, 393)

Quine, in other words, seems to switch from what I have called a wide-scoped holism to a more modest view.[48] In fact, Quine even seems to give up on the idea that "no statement is immune to revision" in the later stages of his career. Whereas the early Quine held that we could amend "statements of the kind called logical laws" in the light of adverse experience (TDE, 1951a, 43), the later Quine appears to make a substantive exception for logic: even if we tried to revise a logical law like the law of noncontradiction, we would "only [be changing] the subject" (PL, 1970a, 81).[49]

In the remainder of this chapter, I will examine Quine's evolving views on the analytic-synthetic distinction after 1951 and argue that the received view about his later position is misguided. I distinguish between three ways in which Quine's early holism can be said to be wide-scoped (section 6.3.1) and show that he never changed his mind about any one of these aspects of his early view (sections 6.3.2–6.3.4). Instead, I argue that Quine's apparent change of mind can be explained away as a mere shift of emphasis.

[47] See also Quine (FME, 1975a, 71): "When we look thus to a whole theory or system of sentences as the vehicle of empirical meaning, how inclusive should we take this system to be? [. . .] modest chunks suffice, and so may be ascribed their independent empirical meaning."

[48] This reading of the evolution of Quine's position is omnipresent in the literature. See, for example, Massey (2011, p. 256): "Late-Quine rails against *extreme holism* [. . .] while advocating *moderate holism*"; and Loeffler (2005, p. 173): "at least since the mid-1970s Quine had moved away from radical holism. [. . .] According to [Quine's later doctrine of moderate holism] the unit of empirical significance is not an all-encompassing background theory (the 'whole of science') any more."

[49] See also Fogelin (2004, p. 32): "not only does Quine's extreme holism become muted in his later writings, the radical revisability thesis associated with it has become muted as well." Quine's apparent change of mind on the status of logic has inspired Arnold and Shapiro (2007, p. 276) to distinguish between a "radical Quine" and a "logic-friendly Quine."

6.3.1. "The Whole of Science"

In section 4.2, I have argued that Quine's holism can be defined in terms of two theses about the logical relation between theoretical sentences and observation categoricals:

(PT) *Prediction thesis*: a single hypothesis does not imply an observation categorical. Only clusters of theoretical sentences will imply observation categoricals.

(FT) *Falsification thesis*: whenever a predicted observation categorical turns out to be false, one cannot logically determine which theoretical sentence is falsified. Rather, the cluster of theoretical sentences that implied the categorical is falsified as a whole.

In "Two Dogmas," Quine defends a wide-scoped variant of holism: he claims that "*total science*" is like "a field of force whose boundary conditions are experience"; that "*the whole of science*" is "the unit of empirical significance"; and that "*no statement*" is "immune to revision" (TDE, 1951a, 42–43, my emphasis). In 1951, in other words, Quine suggests that holism as defined by PT and FT applies to science "as a whole."

Still, Quine's wide-scoped holism, thus described, is strongly ambiguous. Some of the above-mentioned quotes from "Two Dogmas" deal with the domain of evidential holism, whereas other quotes seem to express Quine's views about revisability and/or what he describes as "the unit accountable to an empiricist critique" (TDE, 1951a, 42). In order to unambiguously answer the question whether Quine changed his mind about the scope of his holism, in other words, we need to distinguish between three ways in which holism might be extended to the whole of science, i.e. between three varieties of "wide-scoped holism": maximal inclusion, universal revisability, and maximal integration.

Maximal inclusion: The strongest form of wide-scoped holism is directly connected to the definitions of PT and FT above. These definitions state that only "clusters" of theoretical sentences imply observation categoricals and that only "clusters" of theoretical sentences are falsified whenever these predicted categoricals turn out to be false. A question that naturally arises is how large these clusters are. Or, as Quine phrases it:

> [H]olism [. . .] says that scientific statements are not separately vulnerable to adverse observation, because it is only jointly as a theory that they imply their observable consequences. [. . .] If it is only jointly as a theory that the scientific statements imply their observable

> consequences, how *inclusive* does that theory have to be? (EESW, 1975c, 228–29, my emphasis).

A wide-scoped holist in the first sense would answer Quine's question by claiming that the theory should be *maximally inclusive*. In terms of PT and FT, only science as a whole "will imply observation categoricals" and "whenever a predicted observation categorical turns out to be false," only science as a whole is falsified. The wide-scoped holist in the *maximal inclusion* sense, in other words, believes that "our scientific system of the world [is] involved *en bloc* in *every* prediction" (FME, 1975a, 71, my emphasis).

Universal revisability: A weaker form of wide-scoped holism is connected to Quine's notorious claim that "no statement is immune to revision." If one believes this to be the core of Quine's wide-scoped holism, one need not ascribe anything as strong as *maximal inclusion* to Quine. After all, universal revisability does not imply that every statement is up for revision whenever a predicted observation categorical turns out to be false. It only implies that every statement is revisable, and hence that every statement is required in implying *at least one* observation categorical. Or, in terms of PT and FT, universal revisability is an answer to the question of how many theoretical sentences are revisable in principle because they are involved in implying *at least one* observation categorical.

In order to grasp the distinction between *maximal inclusion* and *universal revisability*, note that the set of *inclusive* theoretical sentences in the sense specified above is a (nonstrict) subset of the set of *revisable* theoretical sentences. After all, if for every observation categorical, all theoretical sentences are required in implying it (*maximal inclusion*), they will by definition play a role in implying *at least one* observation categorical (*universal revisability*). The converse is not (necessarily) the case, as there might be theoretical sentences that play a role in implying only a few but not all observation categoricals. These theoretical sentences are in principle vulnerable to revision, but not *every* time a predicted categorical proves mistaken.

Maximal integration: A third type of wide-scoped holism is connected to the fact that science, at least institutionally, is divided into a great number of subdisciplines. As stated, holism as defined by PT and FT is a thesis about "the logical relation between theory and evidence." This invites the question what one means by "theory." Quine in "Two Dogmas" speaks about "the totality of our so-called knowledge or beliefs, from the most casual matters of geography and history to the profoundest laws of atomic physics or even of pure mathematics and logic" (TDE, 1951a, 42). In principle, however, it possible to endorse *maximal inclusion* and/or *universal revisability*, while rejecting this picture of science. That is, one could defend a pluralistic picture of science but still hold that *maximal inclusion* and/or *universal revisability* are true for every one of the specific subdisciplines of science. An interpretivist in the philosophy of social science, for instance, could accept that natural and social science are separate enterprises with radically different aims, methods, and theories, yet accept that PT and FT apply to both the natural and the social sciences, even in the wide-scoped sense as defined by *maximal inclusion* and *universal revisability*.[50] Wide-scoped holism in the third sense, or *maximal integration*, rejects such a pluralistic picture and claims that all scientific subdisciplines form an integrated whole.[51]

I have specified three ways in which one can attribute a wide-scoped holism to the early Quine. In "Two Dogmas," Quine defends the claim that "the unit of empirical significance is the whole of science" (*maximal inclusion*), that "no statement is immune to revision" (*universal revisability*), and that both these claims apply to "the totality of our so-called knowledge or beliefs" (*maximal integration*). In the remainder of this chapter, I will argue that Quine never changed his mind about these varieties of wide-scoped holism after "Two Dogmas."

6.3.2. Maximal Integration

Let me start with Quine's views about the domain of PT and FT. Reconstructing the evolution of Quine's ideas about *maximal integration* is not a terribly complicated affair. Quine is well known for his use of metaphors that illustrate how the different subdisciplines of science should be viewed as an integrated whole: besides the "man-made fabric" and "field of force" metaphors in "Two Dogmas," his "web of belief" and "Neurath's boat" metaphors have been particularly influential. And although these images have slightly different connotations, they all suggest that Quine was always strongly committed to *maximal integration*.

This is not to say, however, that Quine refuses to accept that science is compartmentalized in practice, i.e. that he denies that most scientists in their everyday work are predominantly concerned with their own subdisciplines. To the contrary, when Quine explicitly comments on science's "degree of integration" (EESW, 1975c, 229), he accepts that science "is variously jointed, and

[50] To clarify, this interpretivist would defend four theses: two claims about the social sciences (PT_S and FT_S) and two claims about the natural sciences (PT_N and FT_N).

[51] Although *maximal integration* is related to the historically influential "unity of science" thesis, I have refrained from dubbing the third sense of wide-scoped holism "unity of science." The reason is that the phrase "unity of science" has strong ontological connotations. Indeed, when Quine himself writes about the "unity of science," he is mainly talking about an ontological unity, i.e. about the "dream of an overarching, unifying fact of the matter" (NLOM, 1995c, 471). And although Quine endorses this drive "for an unified all-purpose ontology" (p. 471), *maximal integration* as specified above is not necessarily connected to these ontological considerations. That is, one could accept *maximal integration* without accepting ontological unity, for instance, when one believes that the sciences are only *methodologically* unified.

loose in the joints in varying degrees" (p. 230). In practice, Quine maintains, "widely separate areas of science can be assessed and revised independently of each other" (RJV, 1986f, 620).

Still, this "practical compartmentalization" (p. 620) does not affect Quine's view that science is an integrated whole *in principle*. The reason why he believes this becomes clear when we consider the ideas of Penelope Maddy, who has challenged Quine on this issue by defending a form of methodological pluralism. In her *Naturalism in Mathematics*, Maddy ponders the question on what basis mathematicians should accept or reject candidate axioms for set theory and argues that only intracompartmental (in this case, intramathematical) norms should play a role:

> Where Quine holds that science is not answerable to any supra-scientific tribunal, and not in need of any justification beyond observation and the hypothetico-deductive method, [my] naturalist adds that mathematics is not answerable to any extra-mathematical tribunal and not in need of any justification beyond proof and the axiomatic method. Where Quine takes science to be independent of first philosophy, my naturalist takes mathematics to be independent of both first philosophy and natural science [. . .] in short, from any external standard. (Maddy, 1997, p. 184)

The difference between Maddy and Quine shows how diverging views about *maximal integration* can have considerable practical ramifications. Whereas Quine proposes to evaluate candidate axioms for set theory by submitting them "to the considerations of simplicity, economy, and naturalness that contribute to the molding of scientific theories generally" (PT, 1990a, 95), Maddy proposes that we set aside these scientific norms because mathematicians themselves appeal to norms "of a sort quite unlike anything that turns up in the practice of natural science: crudely, the scientist posits only those entities without which she cannot account for our observations, while the set theorist posits as many entities as she can, short of inconsistency" (Maddy, 1997, p. 184).

In the wake of Maddy's book, two questions have dominated debates about her strict distinction between science and mathematics: (1) *why* should mathematics be evaluated on its own terms? and (2) *to what extent* should mathematics be shielded from external influence?[52] Now, it is not my intention here to recap the debate that was spawned by Maddy's book. Still, I think that the *communis opinio* on both questions today shows why Quine believes that science should be viewed as an integrated whole in principle, even if in practice most scientists are

[52] See, for example, Dieterle (1999), Hale (1999), Rosen (1999), Tennant (2000), Tappenden (2001), Decock (2002b), and, for an overview, Paseau (2013).

solely concerned with their own subdisciplines. For in discussing the question why mathematics (and not, for example, astrology) should be evaluated on its own terms, many commentators have pointed out that mathematics deserves this privilege because it is, as Maddy admits, "staggeringly useful, seemingly indispensable, to the practice of natural science, while astrology is not" (Maddy, 1997, p. 204). If mathematics, like astrology, had been a useless discipline, in other words, nobody would have felt the appeal of Maddy's proposal. After all, no serious naturalist would deny that we can and should scientifically evaluate (and dismiss) astrological claims. Still, this very use of mathematics also shows why Maddy's strict distinction between science and mathematics ultimately cannot be maintained, i.e. why Quine believes that science ought to be viewed as an integrated whole in principle. For, as many commentators have argued, even if Maddy is right that mathematical decisions are solely made on the basis of intramathematical considerations *in everyday practice*,[53] it is mathematics' use in the scientific enterprise as a whole that *ultimately* justifies why we consider it to be such a tremendously valuable enterprise in the first place, meaning that, when push comes to shove, scientific norms should also play some role in mathematical decisions:

> [A]s Maddy herself points out [. . .] mathematics is "staggeringly useful [. . .] to the practice of natural science." One might add also: to engineering and technology; to medical diagnostics; to the financial markets; to actuarial science; and to a host of other areas of human activity in which everyone's interests and concerns are engaged. So [. . .] those outside the community of professional mathematicians have a permanent and legitimate concern in the nature of the norms governing the latter's practice. In a word: mathematicians cannot be allowed to be a law unto themselves. What they do is too important, and ought to be subject to outside constraints designed to protect everyone's interests. [. . .] For Quine, it is the evidential holism in our theory of nature that truly naturalizes other areas of thought, such as mathematics. Natural science as a whole has to be understood in its own terms. To be anti-holistic, and separate mathematics off from science as a whole, as Maddy does, is to divert the springs of naturalism at their very source.[54] (Tennant, 2000, pp. 238–39)

[53] Something that can be doubted, as mathematicians sometimes *are* influenced by extramathematical considerations. See Hale (1999, p. 395) and Paseau (2013, §5.1).

[54] See also Dieterle (1999), Rosen (1999), and Tappenden (2001). For an evaluation of Maddy's reply to these arguments, see Verhaegh (2015, ch. 6). It should be noted that defining the value of mathematics in terms of its use in science has consequences for the inapplicable parts of mathematics. Quine vacillated quite a lot on this issue. See, for example, (RCP, 1986d, 400; FSS, 1995a, 56).

Maximal integration, in sum, is not a thesis about scientific practice, but a claim about the ultimate justificatory structure of science; even if in everyday practice mathematical decisions are made on the basis of intramathematical considerations, in periods of crisis mathematicians have to appeal to scientific standards, because it is mathematics' use in science that, in the end, justifies why we consider mathematics to be such an immensely valuable practice. Or, as Quine puts it in *Pursuit of Truth*—this time using one of Wittgenstein's metaphors—science in its entirety defines "a particular language game [...] the game of science" (PT, 1990a, 20). The subdisciplines of science form an integrated whole because they are all ultimately part of the same game, because they all contribute to the same goal, maximizing prediction and understanding. It is because he viewed all the subdisciplines of science as part of this one single game that Quine, throughout his career, cherishes *maximal integration* as a regulative ideal.

6.3.3. Universal Revisability

Let me turn to the second type of wide-scoped holism distinguished above. Whereas Quine remains strongly committed to *maximal integration* after "Two Dogmas," he *does* seem to change his mind about *universal revisability*. Indeed, whereas the early Quine argues that "no statement is immune to revision" and that "[r]evision even of the logical law of the excluded middle has been proposed as a means of simplifying quantum mechanics," in *Pursuit of Truth*, Quine seems to explicitly argue that logical truths are exempted from revision:

> Over-logicizing, we may picture the accommodation of a failed observation categorical as follows. We have before us some set S of purported truths that was found jointly to imply the false categorical. [. . .] Now some one or more of the sentences in S are going to have to be rescinded. *We exempt some members of S from this threat on determining that the fateful implication still holds without their help.* Any purely logical truth is thus exempted, since it adds nothing to what S would logically imply anyway. (PT, 1990a, 14, my emphasis)

Quine, in other words, argues that logical truths are not on a par with ordinary scientific statements. Even if we were to decide to remove a logical truth σ from our theory in the light of an unexpected experimental result, this would be a pointless maneuver. For σ would simply pop up again as it is directly implied by S's underlying logic. No matter how often one decides to rescind σ, one will never get rid of it because it follows from the empty set.[55]

[55] See also Quine (TR, 1994a, 431): "Even mathematical truths share [. . .] in the empirical

Despite appearances to contrary, however, Quine's slightly altered perspective on logical truths here does not imply that logic is unrevisable. For it might still be possible to revise the underlying logic that causes σ to pop up every time; if σ automatically follows from one's underlying logic L, then it might still be possible to revise σ via a revision of L, even though it is impossible to revise σ directly.[56] This is certainly not a concession to *universal revisability*, as similar situations occur in other fields of inquiry.[57]

Nothing in the above-quoted passage thus directly implies that, when it comes to *universal revisability*, Quine makes a substantive exception for logic; indirect revisions of logic remain a possibility. Still, the question whether or not Quine allows such indirect modifications has been a matter of some debate. For Quine sometimes seems to suggest that logical principles cannot be revised even in this indirect way; for example, when he claims that the very idea of someone accepting a sentence of the form "p and not p" is "meaningless" (CLT, 1954a, 109), or when he claims that logic is analytic in that a deviant logician who "tries to deny ['p and not p']" only "changes the subject" (PL, 1970a, 81).[58]

It is sometimes thought that these ideas about logic are a consequence of Quine's late view that it *is* possible—*pace* "Two Dogmas"—to defend a limited notion of analyticity (see section 6.2.6). Central to Quine's early holism was the idea that we cannot strictly distinguish between changes of theory and changes of meaning. Quine's renewed ideas about analyticity, however, partly go against this view in acknowledging that a change of logic *can* be a pure change of meaning. For if "p & q ⊨ p" is analytic in Quine's sense, then to deny this logical law is simply to change the meaning of "and," not to propose a change of theory.

In response to these new Quinean ideas about analyticity, scholars have distinguished between what Arnold and Shapiro (2007) call a "logic-friendly" and a "radical" Quine:

> It is sometimes said that there are two, competing versions of [. . .] Quine's unrelenting empiricism [. . .] [The] *logic-friendly* Quine holds that logical truths and, presumably, logical inferences are analytic in the

meaning of sciences where they are applied. [. . .] This cannot be said of logical truths. Any sentence already implies any logical truth, and thus gains no further implying power by being conjoined with it."

[56] One could, for instance, modify L's underlying consequence relation. See Tamminga and Verhaegh (2013).

[57] E.g. certain low-level empirical laws too will have to be "exempted from revision" if the general laws from which are derived are not modified.

[58] See also Quine (TDR, 1991b, 396): "Anyone who goes counter to *modus ponens*, or who affirms a conjunction and denies one of its components, is simply flouting what he learned in learning to use 'if' and 'and.'"

> traditional sense: they are true solely in virtue of the meaning of the logical terminology. Consequently, logical truths are knowable a priori, and, importantly, they are incorrigible, and so immune from revision. No amount of empirical data can get us to revise them. [. . .] The other, *radical* version of Quine does not exempt logic from the attack on analyticity and a priority. Logical truths and inferences are themselves part of the web of belief, and the same methodology applies to logic as to any other part of the web. [. . .] Everything, including logic, is up for grabs in our struggle for holistic confirmation.[59] (Arnold & Shapiro, 2007, pp. 276–77)

There has been considerable controversy about the extent to which the logic-friendly and the radical Quine are compatible, and if they are not, which one of these characters best approaches the "real Quine."[60] With one exception, however, all scholars (including myself, see Tamminga and Verhaegh, 2013) have overlooked the fact that Quine answered this question explicitly on two occasions.[61] According to Quine, the two perspectives on logic are perfectly compatible because his renewed talk about analyticity does not have any significant epistemological consequences. Even if changes of logic are now viewed as changes of meaning, the fact that logical laws are analytic does not have any implications for Quine's views about the epistemological status of logic and, hence, for *universal revisability*. For, Quine argues, if we are unwilling to treat a logical revision as a change of theory, this only reflects how deeply embedded the laws of logic are in our system of beliefs; the maxim of minimum mutilation still suffices to account for the ground of logical truth.

Quine's point here can be illustrated using his ideas about translation. Consider a lexicographer who aims to translate a native tribe's language into English; and suppose that all members of the tribe are inclined to assent to "q ka bu q," which seems to mean "p and not p" if the linguist would follow the translation manual she has drawn up thus far. Now, in response to this situation, the linguist has at least two options. She can either stick to her earlier conclusions and interpret the natives as accepting contradictions; or she might take the natives' utterances as evidence that her existing translation manual cannot be correct. Now, according to Quine, it would be absurd to choose the former option:

[59] See also Haack (1977), who distinguishes between a "conservative" and a "radical" Quine; and Parent (2008), who contrasts a "Principle of Logical Charity" with a "Revisability Doctrine."

[60] See Haack (1977), Shapiro (2000, p. 334), Burgess (2004), Maddy (2005, p. 443), Weir (2005, p. 463), Arnold and Shapiro (2007), Parent (2008), and Tamminga and Verhaegh (2013).

[61] See Quine (RS1, 1968h; CB, 1990c). The exception I am referring to is Parent (2008). Still, even he takes into account only one of these two papers.

> [I]f any evidence can count against a lexicographer's adoption of 'and' and 'not' as translations of 'ka' and 'bu', certainly the natives' acceptance of 'q ka bu q' as true counts overwhelmingly. We are left with the meaninglessness of the doctrine of there being pre-logical peoples; pre-logicality is a trait injected by bad translators. (CLT, 1954a, 190)

If Quine is right, then we cannot but conclude that the natives agree with us when it comes to logic. The very idea of an empirical observation that would justify the lexicographer in ascribing to them an alternative framework is ruled out from the beginning. Yet the fact that we choose to interpret the natives as agreeing with us when it comes to logic does not tell us anything about the *revisability* of logic. We only interpret the native in terms of our logic because it is a basic *pragmatic* rule to interpret one another as charitably as possible:

> What is interesting to ponder is the connection between this rigidity of logic in translation and the question of the immunity of logic to revision [. . .] generally, we are well advised in translation to choose among our indeterminates in such a way, when we can, that sentences which natives assent to as a matter of course become translated into English sentences that likewise go without saying. This policy is regularly reflected in domestic communication: when our compatriot denies something that would seem to go without saying, we are apt to decide that his idiolect of English deviates on some word. [. . .] We see, then, how it is that "Save logical truth" is both a convention and a wise one. And we see also that it gives logical truths no epistemological status distinct from that of any obvious truths of a so-called factual kind. (RS2, 1968h, 317–18)

Confronted with the objection that his ideas about the relation between logic and translation appear to be in conflict with *universal revisability*, therefore, Quine responds by showing that the two theses can be easily combined. His views concerning the interpretation of deviant logicians (both domestic and abroad) are not intended to express anything fundamental about the epistemological status of logic. Rather, Quine only intends to show that translation practices are constrained by the principle of charity, a principle closely related to the maxim of minimum mutilation in theory revision; in updating our theories and in making sense of one another we are inclined to "save the obvious," nothing more, nothing less. Obviousness is not a trait that is exclusive to logic, nor does the fact that a truth is obvious imply that it cannot be revised: "Obviousness resists change but does not preclude it" (CB, 1990c, 36).[62]

[62] See also Quine (CB, 1990c, 36): "Is change of logic a change of language, or is it a change of substantive theory on a par with changes in physics. [. . .] I have seemed to oscillate between those

6.3.4. Maximal Inclusion

I have shown that Quine's ideas about the *scope* of evidential holism ought to be divided into three subtheses: *maximal inclusion, universal revisability,* and *maximal integration,* and I have argued that although Quine seems to have changed his mind about the latter two claims—revising his ideas about analyticity and admitting that science is compartmentalized in practice—these modifications did not affect his commitment to either *universal revisability* or *maximal integration.* Quine still maintains that "logic is integral to our system of the world and accessible to change in the same way as the rest" (CB, 1990c, 36) as well as that the compartmentalization of science is "a matter of practice rather than principle" (RJV, 1986f, 620).

Now, let me finally turn to *maximal inclusion,* the strongest type of wide-scoped holism Quine defends in "Two Dogmas." Prima facie, it seems obvious that Quine *did* change his mind about this issue. (Re)consider, for example, the following passages:

> Looking back on ["Two Dogmas"], one thing I regret is my needlessly strong statement of holism. [. . .] In later writings I have invoked not the whole of science but chunks of it, clusters of sentences just inclusive enough to have critical semantic mass. By this I mean a cluster sufficient to imply an observable effect of an observable experimental condition. (TDR, 1991b, 393)
>
> When we look thus to a whole theory or system of sentences as the vehicle of empirical meaning, how inclusive should we take this system to be? [. . .] modest chunks suffice, and so may be ascribed their independent empirical meaning. (FME, 1975a, 71)
>
> I see science as a considerably integrated system of the world. [. . .] But we can appreciate this degree of integration and still appreciate how unrealistic it would be to extend a Duhemian holism to the whole of science, taking all science as the unit that is responsible to observation. (EESW, 1975c, 229–30)

positions. But are they really two positions? If someone persists in a simple logical falsehood, we do indeed conclude that he has mislearned our language or is tampering with it. But this is equally the way with any obvious falsehood, logical or not." In response to the above-sketched dilemma of Arnold and Shapiro, we can thus say that the "logic-friendly Quine" and the "radical Quine" are perfectly compatible. Their description of these "two Quines" ought to be adapted, however. For although the logic-friendly Quine argues that logic is analytic, he does not believe that it is "analytic in the traditional sense," nor does he believe that logical truths "are knowable a priori" or that they are "incorrigible, and so immune from revisions."

In all these passages Quine appears to be discussing *maximal inclusion*; and in all these passages Quine seems to retract his "Two Dogmas" claim that "the unit of empirical significance is the whole of science." Rather than claiming that science as a whole is falsified whenever an observation categorical turns out to be incorrect, Quine now seems to maintain that science is not *maximally* inclusive; chunks of theory suffice to imply observation categoricals and, as a result, only chunks of theory are falsified whenever these predictions turn out to be false (in what follows, I will call this adapted thesis *moderate inclusion*).

Yet Quine did *not* change his mind about *maximal inclusion*. Rather, Quine believes that *maximal inclusion* is a strictly true but ultimately uninteresting thesis. That is, although Quine does not believe *maximal inclusion* to be false, he wants to shift the focus toward the aspects of holism that he thinks *are* important. As a start, let us consider how the above three passages continue:

> Looking back on ["Two Dogmas"], one thing I regret is my needlessly strong statement of holism. "The unit of empirical significance is the whole of science." [. . .] *This is true enough in a legalistic sort of way, but it diverts attention from what is more to the point: the varying degrees of proximity to observation.* (TDR, 1991b, 393)

> When we look thus to a whole theory or system of sentences as the vehicle of empirical meaning, how inclusive should we take this system to be? [. . .] *It is an uninteresting legalism* [. . .] *to think of our scientific system of the world as involved* en bloc *in every prediction.* (FME, 1975a, 71)

> [W]e can [. . .] appreciate this degree of integration and still appreciate how unrealistic it would be to extend a Duhemian holism to the whole of science. [. . .] *Little is gained by saying that the unit is in principle the whole of science, however defensible this claim may be in a legalistic way.* (EESW, 1975c, 230).

In all three instances, Quine does not reject *maximal inclusion* because it is false, but because it is somehow unimportant: he claims that "it diverts attention from what is more to the point," that it is "uninteresting," and that "little is gained" in advancing the thesis. Furthermore, instead of claiming that *maximal inclusion* is false, in all three passages Quine claims that the thesis is true (or at least defensible) "legalistically."[63] Quine's point is that although *in practice* relatively small chunks of theory are inclusive enough to imply an observation categorical,

[63] See also Quine's (RHP, 1986e, 427), "I see extreme holism itself as 'pure legalism' "; (PL, 1970a, 5), "Legalistically, one could claim that evidence counts always for or against the total system, however loose-knit, of science"; and (RJV, 1986f, 620), "Holism at its most extreme holds that science faces the tribunal of experience [. . .] as a corporate body. [. . .] Legalistically this again is defensible."

logically science as a whole is involved. After all, even if one deduces an observation categorical from a small chunk of theory, one could still decide to save the sentences in the chunk when the categorical turns out to be false; for example by revising the logic that is shared by all the chunks or by revising the rules of one's language. In other words, although relatively small chunks of theory will imply observation categoricals *given* a certain language and a certain logic, Quine believes it to be legalistically true that only science as a whole implies observation categoricals when one takes into account the scientist's freedom to revise her logic and/or her language in the light of adverse experience. And although it is trivially true that one can revise anything whatsoever if one is willing to modify the rules of one's language, this is exactly why Quine believes *maximal inclusion* to be "true in a legalistic sort of way."[64]

So Quine does not believe *maximal inclusion* to be false. A question that remains to be answered, however, is why he so strongly emphasizes *moderate inclusion* (his chunk-view) in later work. It is my contention that Quine backs away from *maximal inclusion* in later stages of his career (i.e. from *Word and Object* onward),[65] because it is only then that he starts to realize that *maximal inclusion* does not do any epistemological work in his philosophy. Rather, Quine realizes, *universal revisability* and *moderate inclusion* suffice.

To see this, reconsider Quine's second argument in "Two Dogmas of Empiricism." We have seen that *if* Quine's holistic picture of inquiry is correct, "it becomes folly to seek a boundary between synthetic statements, which hold contingently on experience, and analytic statements which hold come what may." This is folly, according to Quine, because on the holistic picture "Any statement can be held true come what may" and, conversely, "No statement is immune to revision" (TDE, 1951a, 43). Now, for Quine's argument here to be valid, he does not require *maximal inclusion*; his more modest "chunk-view" is sufficient to establish that one can hold any statement true come what may. After all, if a relatively small chunk of theoretical sentences is always involved in implying an

[64] See also Quine's response to Grünbaum, who makes a related point: "Your claim that the Duhem-Quine thesis, as you call it, is untenable if taken nontrivially, strikes me as persuasive. [. . .] For my own part I would say that the thesis as I have used it *is* probably trivial. I haven't advanced it as an interesting thesis as such" (CGC, 1962, 132). Note that Quine has not changed his mind about this issue either. Already in the first edition of *Methods of Logic*, Quine recognizes that some radical revisions can better be thought of as "the adoption of a new conceptual scheme, the imposition of new meanings on old words" (ML1, 1950a, xiv).

[65] See (WO, 1960, 12–13): "In obvious way this structure of interconnected sentences is a single connected fabric including all sciences, and indeed everything we ever say about the world; for the logical truths at least, and no doubt many more commonplace sentences too, are germane to all topics and thus provide connections. However, some middle-sized scrap of theory usually will embody all the connections that are likely to affect our adjudication of a given sentence."

observation categorical, one has the logical freedom to decide never to revise a particular hypothesis; there will always be at least some auxiliary hypotheses that can be blamed whenever a prediction turns out to be false. Quine *does* need *universal revisability*, however, if he wants to maintain that "no statement is immune to revision." For if Quine would moderate his view about revisability, he could obviously not maintain that any statement is revisable in principle.

Quine in other words, only needs *universal revisability* and *moderate inclusion* in arguing that we do not need an analytic-synthetic distinction to account for logical and mathematical knowledge; *maximal inclusion* is not required.[66] In order to show that mathematical and logical truths play an important role in squaring theory with evidence, Quine does not need to maintain that they are involved in implying each and every observation categorical. If they are involved in implying at least a few categoricals, and hence are revisable in principle, he can legitimately claim that we do not require analyticity in order to justify our logical and mathematical knowledge.[67]

Although Quine believes *maximal inclusion* to be true legalistically, in other words, he has no reason to advance the thesis as such. Both dogmas Quine wants to reject turn out to be false if weaker versions of his wide-scoped holism—viz. *moderate inclusion* and *universal revisability*—are true. Just as Quine never changed his mind about *maximal integration* and *universal revisability*, in sum, he never changed his mind about *maximal inclusion* either; he just literally believes it to be "needlessly strong."

6.4. Conclusion

In this chapter, I have reconstructed Quine's evolving views on the analytic-synthetic distinction. I have argued that it is impossible to identify a specific moment at which Quine came to fully reject the distinction, not only because Quine never definitively abandoned the search for a behavioristically acceptable definition of analyticity, but also because his views were subject to a number of changes in the first decades of his career. I have argued, for instance, that Quine challenged Carnap's semantic version of the analytic-synthetic distinction in

[66] See Quine's introduction to the third edition of *From a Logical Point of View* (which, unfortunately, never appeared in print): "The holism in "Two dogmas" has put many readers off, but I think its fault is one of *emphasis*. All we really *need* in the way of holism for the purposes to which it is put in that essay, is to appreciate that empirical content is shared by the statements of science in clusters. [. . .] *Practically* the relevant cluster is indeed never the whole of science (September 1979, item 2836, my transcription and emphasis).

[67] See Quine (QSM, 1988, 27): "Once we appreciate holism, *even moderate holism*, the notion of analyticity ceases to be vital to epistemology."

1940, that he became aware of the nature of their disagreement in 1943, and that he adopted his wide-scoped holism in 1948. Sometimes these developments were fueled by like-minded philosophers (e.g. Tarski and, to a lesser extent, Goodman), and sometimes these changes were stimulated by developments in other domains (e.g. his "rapidly evolving" views about the epistemic status of mathematical objects). After "Two Dogmas," however, Quine's views on the analytic-synthetic distinction remained relatively stable. Although Quine came to embrace a "vegetarian notion of analyticity" in *Word and Object* and started to emphasize that we do not *need* to presuppose that science as a whole is at stake in every prediction, he never changed his mind about the analytic-synthetic distinction itself: there simply is no definition of analyticity that is both behavioristically acceptable *and* can play the role that Carnap believed the analytic-synthetic distinction should play.

7

Science and Philosophy

7.1. Introduction

In contemporary philosophy, naturalism is often defined as a *metaphilosophical* thesis. According to contemporary naturalists, we cannot study the nature of mind, knowledge, language, meaning, and reality without taking into account the results from physics, linguistics, and the cognitive sciences. Any metaphysical theory about time has to incorporate special relativity, and any theory about consciousness has to be compatible with our best neuroscientific theories. Although there seem to be as many variants of naturalism as there are naturalists, there seems to be widespread agreement that philosophy ought to be at least *continuous with* science.

In the previous chapters, I have distinguished two Quinean arguments for metaphilosophical naturalism. According to Quine, there cannot be a strict distinction between science and philosophy because (1) there are no transcendental, distinctively philosophical perspectives on science and because (2) we cannot draw a strict distinction between matters of fact and matters of language—between the analytic and the synthetic. To be a Quinean naturalist, we have seen, is recognize that we are always working from within our inherited theory as a going concern.

In the early 1950s, we have seen, Quine had both arguments at his disposal. He had conclusively dismissed the problem of epistemic priority (chapter 5) and had accepted that there is no epistemologically significant analytic-synthetic distinction (chapter 6). In this final historical chapter, I will examine Quine's evolving views on the science-philosophy distinction. Building on the material available at the W. V. Quine Papers, I piece together the development and reception of Quine's metaphilosophy between 1953, when he wrote the first draft of *Word and Object*, and 1968, when he adopted the label "naturalism" for his philosophy.

This chapter will be structured as follows. I start with a reconstruction of Quine's metaphilosophical views in 1953 by examining, among other sources, a previously unexamined 454-page transcript of his first working draft of *Word and Object*. I argue that Quine's metaphilosophical views at the time were still strongly influenced by the ongoing debate between positivists and ordinary language philosophers (section 7.2). Next, I show that Quine became increasingly interested in psychology and linguistics in the years leading up to the publication of *Word and Object* (1960) and argue that this development affected his perspective on the relation between science and philosophy (section 7.3). In the final two sections, I reconstruct the way in which Quine's naturalism was received by the philosophical community. I show that Quine was dissatisfied with the responses to *Word and Object* (section 7.4) and argue that these responses contributed to his decision to rebrand his philosophy in the late 1960s, when he for the time adopted the label "naturalism" for his metaphilosophical views (section 7.5).

7.2. Ordinary Language and the Language of Science

More than half a century ago, W. V. Quine published *Word and Object*, one of the most influential monographs of twentieth-century philosophy. In the book—published as part of the Studies in Communication series at MIT Press—Quine presents his seminal views on language, science, and ontology and incorporates them for the first time in a comprehensive naturalistic framework, a metaphilosophy in which all forms of inquiry—philosophy, science, and common sense—are viewed as part of a single continuous enterprise.

Quine, as I have shown in chapter 5, started working on *Word and Object* in 1952, when he had solved the problems that led him to postpone *Sign and Object* in the mid-1940s. In a letter to Roman Jakobson, Quine envisions a book project which portrays the "development of science" as "a gradual warping and adjusting of the pattern of language, or myth, in such ways as seem increasingly to serve the pragmatic purpose of anticipating experience" (May 18, 1952, item 1488).

In order to obtain a first draft for his projected book, Quine redesigned his Philosophy of Language course for the 1953 spring semester and had the twenty-three lectures taped and transcribed by Alice Koller.[1] The resulting 454-page typescript with autograph corrections by Quine (item 3158) almost completely

[1] See Quine's letters to Roman Jakobson: "I am now busily working out a tentative organization of the book, which is at the same time a reorganization of my course in the Philosophy of Language for the coming term" (January 22, 1953, item 1488) and to Paul Buck: "During the past term the lectures in my course on the Philosophy of Language were tape-recorded and typed, with a view to obtaining a working draft for a book in Jakobson's projected series" (June 8, 1953, item 479).

covers the lecture series and can be crosschecked with the approximately 400 three-by-five inch notecards Quine wrote in preparing the course (item 3266). Combined, these sources provide a unique insight into Quine's metaphilosophy in the early 1950s.[2]

For the purposes of this chapter, one of the most interesting aspects of the typescript is the light it sheds on Quine's views on the nature of philosophical questions in the early 1950s. Given his rejection of the analytic-synthetic distinction and his reconception of epistemology as an empirical science, Quine had all the ingredients to defend a thoroughly naturalistic metaphilosophy, i.e. to emphasize the continuity of science and philosophy against the strict distinction presupposed by both the positivists and the ordinary language philosophers at Oxford. The typescript shows, however, that Quine, *like* the positivists and the Oxonians, was primarily concerned with the relation between science and *ordinary language.* The bulk of the typescript deals with the development of a scientific language—i.e. an extensional, nonambiguous, nonvague language—and emphasizes that ordinary language is not "superseded by" but merely "modified for" scientific purposes. Where one would expect Quine to blur the boundary between science and *philosophy*—thereby further developing his rejection of both the analytic-synthetic distinction and the divide between epistemology and psychology—he in fact focuses on the continuity between science and *ordinary language.*

To give an example: when Quine shows how we, in developing a scientific language, can get rid of egocentric particulars (Russell's term for indexicals like "here," "now," and "I") by resorting to names and descriptions of time and place—e.g. "Julius Caesar crossed the Rubicon on January 10, 49 BC" instead of "He crossed this river two thousand years ago"—he emphasizes that although the resulting language does not make use of egocentric particulars, it still *depends* on them in that we need them in order to *understand* these names and descriptions:

> Ordinary language, with its egocentric particulars, is not superseded as dispensable. For example, ostension, which is certainly in the spirit of the egocentric particulars 'here', 'now', 'there' whether or not they are uttered, is needed to establish names for places and dates to take the place of subsequent use of egocentric particulars. Even if we use latitude and longitude to name places, we have still to point to the meridian.

[2] Quine also lectured on the philosophy of language at Oxford during Michaelmas Term in the 1953–1954 academic year, when he visited Oxford as Eastman Professor. Unfortunately, Quine's notes for these lectures seem to have been lost. The collection of note cards at Houghton contains only his notes on logic and set theory for Hilary and Trinity Term (item 3277).

> [. . .] Scientific language is genuinely a splinter or offshoot instead of a replacement of ordinary language.[3] (February 12, 1953, item 3158)

Similarly, when Quine introduces the "logical grammar" of his scientific language, he emphasizes that his grammar does not do anything that cannot already be done in ordinary language:

> For instance, here we have a statement of ordinary language 'All dogs bark' and we try to put it over into logical form [using] universal quantification [$\forall(x)(\neg(Dog(x) \wedge \neg Barks(x)))$]. We can put that over into logical symbols completely now, except for 'barks' and 'is a dog' which are extra-logical components. [. . .] But at the same time, this is—one might say—English [. . .] each of these signs can be given a fairly direct [. . .] paraphrase into ordinary language. 'Everything is such that it is not the case that it both is a dog and it is not the case that it barks.' And it's only because of *that* fact that we can understand these artificial symbols. So what's happened isn't that we've put ordinary language into a very un-English sort of symbolism, but rather that we're [. . .] paraphrasing ordinary language into a part of ordinary language. (March 3, 1953, item 3158, my emphasis)

The purpose of this paraphrasing, of course, is that the subset of ordinary language into which ordinary language as a whole is paraphrased is clearer and, hence, more useful for the sciences.

In his first draft of *Word and Object*, in other words, Quine is strongly concerned with the relation between ordinary language and the language of science.[4] This becomes even clearer when he explicitly discusses the relation between his project and the traditional problems of philosophy. Rather than arguing that we can solve philosophical problems by adopting a broadly scientific methodology—e.g. solving the problems of epistemology by "scientifically investigating man's acquisition of science" (RR, 1973, 3)—Quine adopts the attitude that philosophical problems can be *eliminated* by the *analysis of language*,

[3] See Quine (SLS, 1954b, 236) for a published version of this argument: "Terms which are primitive or irreducible, from the point of view of [. . .] scientific notation, may still be intelligible to us only through explanations in an ordinary language rife with indicator words, tense, and ambiguity. Scientific language is in any event a splinter of ordinary language, not a substitute."

[4] See also Quine's description of the book in a letter to Max Millikan: "The book on which I am working for Professor Jakobson's series has the tentative title *Language and Knowledge*. [. . .] One large part of the book will be devoted to logical grammar, conceived as an artificial development of part of ordinary language for increasing utility in its scientific applications" (January 20, 1953, item 1488). In his letter to Jakobson, Quine writes: "Within the matrix of natural language a scientific sublanguage tends to crystallize out, but this crystalline part is even today a small fraction of our language" (May 18, 1952, item 1488).

an attitude that was fashionable at the time, as both the logical positivists and the ordinary language philosophers expressed similar views.

> The philosophical importance of this sort of analysis [. . .] is that they *eliminate*—from time to time—certain philosophical perplexities. And they can be looked upon as eliminating certain philosophical perplexities by showing that the part of language which gave rise to those perplexities is avoidable in favor of an unperplexing part of language. And this, I think, is pretty characteristic of avoidance of philosophical perplexities. (April 16, 1953, item 3158)

According to Quine, in other words, many philosophical problems are linguistic in nature. Just as we can avoid ambiguity and imprecision by paraphrasing ordinary language into a subset of ordinary language, we can eliminate philosophical perplexities by clarifying ordinary language. A few pages later, Quine explicitly connects this metaphilosophy to the approach of the ordinary language philosophers:

> Once you've shown how to avoid these troublesome terms, we've shown that whatever problems exist over them are purely verbal. [. . .] And a verbal problem is, philosophically speaking, no problem at all. It's examples of this sort that led Wittgenstein to say that the task of philosophy is not to solve problems, but to eliminate them, by showing that there were no real problems in the first place. (April 16, 1953, item 3158)

Of course we should not conclude that Quine's metaphilosophy coincides with the view of the ordinary language philosophers. After all, Quine, like the logical positivists, emphasized the role of logical analysis and explication in eliminating philosophical problems. Where Strawson (1950), for example, rejects Russell's theory of descriptions because it conflicts with ordinary language, Quine accepts the theory because it is certainly one very effective way of clearing up "perplexities" about the status of nonexistent entities.[5] Still Quine,

[5] See "Mr. Strawson on Logical Theory," where Quine argues that although logical language "has its roots in ordinary language," it can contribute to the "elimination of philosophical perplexities" (MSLT, 1953b, 150–51). Quine considered this paper as "a manifesto" to "herald" his academic year at Oxford as Eastman Professor (TML, 1985a, 235). See also Quine's 1951 letter to Burton Dreben (who was visiting Oxford at the time): "I liked Strawson's paper. [. . .] I think his objections to Russell's theory of descriptions as a representation of ordinary language are well taken. [. . .] Russell's description theory [is] a simplificatory *departure* from ordinary usage. [. . .] I think Strawson and other Oxonians underestimate the scientific utility of such simplificatory departures, but they are right in insisting that they are departures and in criticizing Russell for failing to see them as such" (January 4, 1951, item 315).

unlike the logical positivists, sides with the Oxonians in arguing that useful artificial languages always *depend on* ordinary language, i.e. that in improving a language we are always working from within. Already in 1943, we have seen, Quine recognized that his "attitude toward 'formal' languages is very different from Carnap's." In a letter to Alonzo Church, Quine argues that he, unlike Carnap, considers "[s]erious artificial notations, e.g. in mathematics or in [. . .] logic [as] supplementary but integral parts of natural language (August 14, 1943, item 224).[6]

In his 1953 draft of *Word and Object*, in sum, Quine seems to adopt a position somewhere between the positivists and the ordinary language philosophers: he maintains that we can eliminate philosophical problems using the tools of modern logic, while emphasizing that these logical tools are grounded in ordinary language:

> The development of science in the past few millennia is a gradual warping and adjusting of the pattern of language [. . .] *Within* the matrix of natural language a scientific sublanguage tends to crystallize out [. . .] a small fraction of our language. Value terms, egocentric particulars, and modalities are foreign to it. The scientific sublanguage is the proper sphere of logic, and of ontological commitment, and of explication, or so-called analysis in the current British sense. (May 18, 1952, item 1488, my emphasis)

Whatever his exact attitude toward the logical positivists and the ordinary language philosophers, however, Quine largely accepts their shared presupposition that most philosophical problems are linguistic problems. Even Quine's reconception of epistemology as an empirical science (see section 5.9) largely depends on considerations about (ordinary) language. In summarizing his argument against sense data in "The Place of a Theory of Evidence," for example, Quine writes:

> We began by observing with regard to molecules that the physicist has no evidence of their existence, beyond the fact that by positing them he can smooth his laws. Next we observed with regard to the external objects of common sense that there is no evidence of their existence beyond the fact that by positing them one get a manageable degree of conceptual organization into his train of experience. Thereupon we made the wrong turning, by concluding that there was no evidence for the existence of any of these objects. What is evidence [. . .] if the testimony of my senses is not evidence of the presence of my desk? [. . .]

[6] See sections 4.7 and 6.2.4.

> To withhold *the name of evidence* from such instances is to warp *the term 'evidence'* away from just the denotations which had originally done most to invest the term with whatever intelligibility it may have for us. (October 7, 1952, item 3011, p. 22, my emphasis)

Quine's argument against sense data, in other words, largely depends on his considerations about the way in which we use the term "evidence." See also "The Scope and Language of Science": "We cannot significantly question the reality of the external world, or deny that there is evidence of external objects [. . .] for [. . .] to do so is simply to dissociate the terms 'reality' and 'evidence' from the very applications which originally did most to invest those terms with whatever intelligibility they may have for us" (SLS, 1954b, 229).

The fact that Quine almost exclusively relies on linguistic considerations in dealing with philosophical problems is surprising. For both the positivist's distinction between "science" and the "logic of science" (Carnap, 1934, §72) and the Oxonian distinction between "talk[ing] sense with concepts" and "talk[ing] sense about them" (Ryle, 1949, p. 7) rely on the analytic-synthetic distinction Quine had given up on a few years before. Where the positivists and the ordinary language philosophers were forced to think of philosophical problems as linguistic problems because of their views about the nature of philosophy, Quine, in rejecting the underlying analytic-synthetic distinction, had the tools for dispensing with this narrow conception of philosophical problems. Quine's first draft of *Word and Object*, however, was still largely compatible with the traditional view.[7]

7.3. Philosophy and Science; Science and Philosophy

In his 1952 letter to Jakobson, Quine projected that he would finish his book in 1954. Still, it would take him almost seven years to complete the book after he had obtained his first working draft; an excessively long period by Quine's standards.[8] During these years, Quine was primarily concerned with the

[7] In his autobiography, Quine writes that he was not satisfied with his first draft. "I was struggling to find the right structure for my work in progress, what was to become *Word and Object*. [. . .] My course at Harvard [. . .] had been recorded on tape and transcribed [. . .] but I set that all aside, lest my quest for the right structure be obstructed by excessive detail. Better to let the transcript lie for a year or two and then mine it for supplementary substance when the structure was in hand. I even dreaded broaching it, what with my spoken hems and haws, my false starts and infelicities" (TML, 1985a, 241–42).

[8] Most monographs Quine wrote were completed within one or two years. Quine's Portuguese book, *O Sentido da Nova Logica* (1943a), was even written in less than four months (TML, 1985a,

development of his positive epistemology (see section 5.9), i.e. his genetic study of "the ways of knowing" by studying "the learning of language" and "the acquisition of scientific concepts" (April 9, 1953, item 3158). Where Quine's 1953 lectures still present a sketchy story that focuses solely on ostension and internal similarity standards, *Word and Object* provides a detailed and complex account involving ocular irradiation patterns, stimulus meanings, babbling, conditioning, mimicry, prelinguistic quality spaces, phonetic norms, discrimination thresholds, collateral information, degrees of observationality, and observation sentences.[9] Perhaps Quine's increased interest in child development was partly prompted by his son Douglas (born in 1950) and his daughter Margaret (born in 1954). Indeed, Quine's autobiography contains a chapter with anecdotes about their questions, phrases, and grammatical mistakes, showing that he kept extensive notes about their linguistic development (TML, 1985a, 273–79).

In developing his genetic account, Quine came to put more and more emphasis on the integration of science and philosophy. In 1956, Quine writes McGeorge Bundy that his work "seems to be turning increasingly relevant to behavioral studies" (December 3, item 205) and Quine also visited one of the interdisciplinary symposia organized by Jean Piaget's Institute for Genetic Epistemology. Piaget even proposed to write a book with Quine entitled *Formal Logic and Real Thinking* (*La logique formelle et la pensée réelle*). In a letter to Evert Willem Beth, Quine writes that he does not agree with Piaget's views on logic, but that the latter's "conceptions and experiments in the field of psychology are really interesting," explaining that he himself has also "been active in the domain of theoretical psychology [. . .] for several years" (March 3, 1960, item 96). Indeed, in a letter to Leo Apostel (who studied with Piaget at the Institute for Genetic Epistemology) Quine writes that "experimental psychology is the proper domain for the analyticity debate" (December 9, 1957, item 40).[10]

159–74). It is therefore no surprise that Quine, in 1959, speaks about his book as "evolving painfully toward completion for nine years and more" (March 2, 1959, item 473).

[9] Although the story of the development of Quine's genetic account is a fascinating one—a story that certainly does not end with *Word and Object*, as large parts of his work after 1960 aim to extend and refine the account provided in his earlier work—I will not go into any details here. For the purposes of this chapter, it suffices to show that Quine, in these years, became "more consciously and explicitly naturalistic" by "turning to our physical interface with the external world" (TDR, 1991b, 398). One of the biggest breakthroughs was Quine's criterion of observationality. In a letter to Davidson, Quine writes that he views this criterion as "a triumph of the Neurath attitude, the embedding of epistemology in natural science in the spirit of his Schiffsumbau simile" (August 13, 1965, item 287).

[10] See also Quine's travel grant application with the Harvard psychologist Jerome Bruner (March 26, 1956, item 475).

In the 1958–1959 academic year, Quine spent two semesters at the Stanford Center for Advanced Study in the Behavioral Sciences, critically discussing his work with "receptively minded linguists, anthropologists, and psychologists" (October 16, 1959, item 205).[11] And although Quine reports that "[n]o appreciable change of theory came out of" these discussions, his turn to the sciences seems to have changed his metaphilosophical perspective. Indeed, in a 1959 summary for Wiley, Quine writes that one of the distinguishing features of *Word and Object* is the fact that "[o]ther logical semanticists put no such weight as I do upon natural linguistics and psychology" (October 17, 1959, item 1488).[12]

It is not my intention here to suggest that Quine, in turning to the sciences, aimed to provide a thoroughly experimental account of language learning in *Word and Object.* Because Quine accepts that "[d]ifferent persons growing up in the same language are like different bushes trimmed and trained to take the shape of identical elephants" (WO, 1960, 8), he is not really interested in "[w]hatever our colleagues in the laboratory may discover of the inner mechanism[s] of [the learning process]" (SLS, 1954b, 231). Indeed, Burton Dreben seems to have advised Quine to speculate as little as possible about underlying psychological mechanisms. In a note titled "Revidenda (Burt, July 16, 1958)," Quine writes: "'Conditioning'; go easy; controversial; Chomsky vs. Skinner. Neutralize the assumption of a specific mechanism as much as possible. Talk of learning, habit formation, etc." (item 3170).[13] Rather, Quine's account is scientific on a more abstract level: he is interested in how these different bushes can be shaped into identical elephants in the first place, considering the fact that all we have to go on are "the effects which [physical objects] help to induce at our sensory surfaces" (WO, 1960, 1).[14]

[11] Among others, Roman Jakobson, Alf Sommerfelt, Gene Galanter, George Miller, and Charles Osgood.

[12] This was also recognized by some readers of *Word and Object.* Cohen (1962, p. 29) writes that "though lexicographers, grammarians and historians of ideas presumably use the concept of meaning at least as often, carefully and knowledgeably as anyone else, modern philosophers propounding a theory of meaning have rarely paid any explicit attention to their use of it. W. V. Quine is perhaps the only exception among philosophers of first rank." Murphey (2012, p. 126) writes that critics of *Word and Object* "raised [the objection] that it was psychology rather than philosophy."

[13] Cf. (WO, 1960, 80–83).

[14] This attitude was shared by the members of Piaget's Institute for Genetic Epistemology. In response to *Word and Object,* Jean-Blaise Grize, for instance, writes: "the subject of the psychology necessary to epistemology is of a very particular nature. [. . .] The behavior of some particular child, with his own set of characteristics, with his personal history [. . .] cannot in any way enlighten the epistemologist. [. . .] It would be infinitely more useful to interview one man such as Prof. Quine than ten thousand readers of *Life* magazine" (Grize, 1965, p. 464).

Nor is it my intention to suggest that Quine completely changed his mind between 1953 and 1960. Quine's mature position is still compatible with the view that philosophical problems are linguistic problems.[15] Unlike the Oxonian and positivist proponents of the linguistic turn, Quine's naturalism is compatible with both the view that philosophical problems are linguistic problems *and* the view that philosophical problems are empirical problems. In fact, Quine's views about analyticity imply that there is no such distinction; an explication is *both* a linguistic proposal and a scientific choice.

What I do want to suggest is that Quine's perspective on the science-philosophy distinction shifted in the years between his first draft and the eventual publication of *Word and Object.* If one, like Quine, gives up on the science-philosophy distinction, there are basically two ways to advance one's view: one can either argue that science is more like philosophy than people used to believe or argue that philosophy is more like science than has often been proposed. In other words, one can emphasize either the role of linguistic and pragmatic factors in science or the role of factual or empirical considerations in philosophy. Now, in the early 1950s Quine clearly takes the former route:

> Carnap maintains that ontological questions, and likewise questions of logical or mathematical principle, are questions not of fact but of choosing a convenient conceptual scheme or framework for science; and with this I agree only if the same be conceded for every scientific hypothesis. (CVO, 1951c, 211)
>
> [W]hat I reject in Carnap's doctrine is not the identification of philosophical tenets with linguistic framework, but the failure to extend that identification indefinitely far into science itself. Thus the relevance of linguistic considerations is, if anything, even more pervasive for me than for Carnap. (February 3, 1953, item 3158)
>
> I grant that one's hypothesis as to what there is [. . .] is at bottom just as arbitrary or pragmatic as one's adoption of a new brand of set theory or a new system of bookkeeping. [. . .] But what impresses me more than it does Carnap is how well this whole attitude is suited also to the theoretical hypotheses of natural science itself. (CLT, 1954a, 132)

Rather than claiming that philosophy is more like science than Carnap used to think, Quine emphasizes that the former's ideas about philosophy also apply to the sciences; science too partly depends on pragmatic criteria. Hence Quine's

[15] Indeed, Quine still connects his metaphilosophy to Wittgenstein's in *Word and Object,* arguing that Wittgenstein's metaphilosophy "aptly fits explication" even if it "has its limitations" (WO, 1960, 260).

famous remark that he is espousing "a more thorough pragmatism" than Carnap and Lewis (1951a, 46).[16]

In the years between 1953 and 1960, however, Quine's perspective shifts to the other side of the coin; i.e. he starts to argue that philosophy is more like science than the proponents of the linguistic turn have advanced.

> The same motives that impel scientists to seek ever simpler and clearer theories adequate to the subject matter of their special sciences are motives for simplification and clarification of the broader framework shared by all the sciences. (WO, 1960, 161)

Quine's regimentation project in the last three chapters of *Word and Object* is highly similar to his development of a scientific language in the 1953 draft. Still, *Word and Object, unlike* the first draft, continuously emphasizes that his regimentation project is "continuous with science in [. . .] motivation" (p. 229), that the "quest of a simplest, clearest overall pattern of canonical notation is not to be distinguished from a quest of ultimate categories (p. 161), and that the only thing that distinguishes his ontological project from the scientist's project is "breadth of categories" (p. 275), clarifications that are largely missing in his first draft. In *Word and Object,* both the first and the last section of his two chapters on regimentation as well as the last section of his chapter on ontology (i.e. §33, §47, and §56) emphasize that his aims are scientific and that it is misleading to think that observation has no "bearing on logic and philosophy" (p. 274). In *Word and Object,* in other words, Quine explicitly adopts the view that philosophy is more like science than philosophers used to believe.

7.4. The Reception of Quine's Naturalism

Word and Object was a big success in terms of attention and sales records.[17] Philosophers who sympathized with Quine's perspective considered the book to

[16] Although Quine, in retrospect, always denied that he defended a variant of pragmatism in the early 1950s ("I was merely taking the word ['pragmatism'] from Carnap and handing it back"), there is quite some evidence that Quine did think of himself as a pragmatist at the time (TDR, 1991b, 397). In the "Present State of Empiricism," written in May 1950 (appendix A.5), Quine likens his view to James's, and argues "for an increased sympathy with the pragmatists" (May 26, 1951, item 3015, my transcription). See also (LDHP, 1946c, 135) and (IOH, 1950b, 78–79). Quine's folder with notes from the early 1950s is even called "Pragmatism, etc." (item 3184). In *Word and Object,* however, Quine abandoned the label, perhaps because it became increasingly unclear to him "what it takes to be a pragmatist" (PPE, 1975b, 23).

[17] By the end of 1961, Quine had sold 3,216 copies. The number of sales strongly increased after the release of a paperback edition in 1964; 23,784 copies of *Word and Object* were sold by 1970 (item 1488). In a 1968 letter to Leigh Cauman, once his PhD student, Quine cheerfully reports

be an absolute masterpiece. In a letter to Quine, John Myhill writes that he rates *Word and Object* "perhaps highest of any philosophical work of this century" (August 30, 1961, item 755) and Peter Hare reports that "some believe *Word and Object* to be one of the most important works of epistemology and metaphysics ever written in [the United States]" (1968, p. 272).[18] Quine's more explicitly naturalistic perspective was received positively as well. J. J. C. Smart, for example, writes that *Word and Object* has "well converted" him to Quine's "philosophical methodology"; he argues that he, "in the vein of *Word and Object*," is "more inclined to say that you can't sharply separate philosophy and science" (July 15, 1960, item 1005).

Outside Quine's closest philosophical circle, however, the responses to *Word and Object* were more varied. Many reviews that appeared in the years immediately following its publication complain about Quine's style, arguing that it is "very demanding" (Wells, 1961, p. 696), too "casual" (Becker, 1961, p. 238, my translation), and "sometimes condensed to the point of enigma" (Presley, 1961, p. 175).[19] It is perhaps because of this reason that some reviewers judge Quine's position to be "unclear" (Oesterle, 1961) and nearly impossible to summarize (Becker 1961), or, even stronger, that the book makes "you wonder what Quine's philosophical position really is" (Galanter, 1961).

This confusion about *Word and Object*'s main message outside Quine's closest philosophical circle had its effect on the reception of his views on metaphilosophy as well. Although some commentators clearly understood the book's main metaphilosophical message,[20] a wide range of positions were ascribed to Quine in the 1960s and early 1970s. Some reviewers interpret Quine's position to be "a version of pragmatism" (Wells, 1961, p. 696), to represent a "pragmatic turn" within analytic philosophy (Smith, 1969, p. 601), or to be a combination "of both Positivism and Pragmatism" (Ammerman, 1965, p. 9); some still seem to view

that his "books sell like hot cakes" (October 14, 1968, item 1422). *Word and Object* was reviewed in philosophical, psychological, and mathematical journals by, among others, Rescher (1960), Presley (1961), Becker (1961), Galanter (1961), Wells (1961), Yourgrau (1961), Oesterle (1961), Goodstein (1961), Geach (1961), Johnstone (1961), Deledalle (1962), and Lieb (1962).

[18] See also Brody (1971, p. 167), "Quine is clearly one of the most important philosophers of our century"; and Davidson and Hintikka (1968, p. 1), "In the philosophical literature of the last decade few if any works contain as great a wealth of ideas [as] *Word and Object*."

[19] In explaining his complaint, Rulon Wells quips: "if the ship of science is not merely to stay afloat but to sail, it should not drag anchor" (1961, p. 696). In his intellectual autobiography, Quine admits that he condensed some sections of *Word and Object* "so as not to harp unduly on the obvious" and that Burton Dreben told him that he "overdid those precautions" (IA, 1986a, 34, 44).

[20] See, for example, Passmore (1966), who in the second edition of *A Hundred Years of Philosophy* (1966) recognizes that Quine "wholly rejects that sharp contrast between philosophy and science which had been common ground to phenomenology, to logical positivism, to Wittgenstein, and to Oxford's 'ordinary language philosophy'" (pp. 531–32).

Quine as a paradigmatic linguistic philosopher (Urmson, 1961; Rorty, 1967b); some claim that "Quine's meta-philosophic theory commits him to a phenomenalism where 'the world' is always 'the world for some subject' " (Yolton, 1967, p. 156); and some read Quine as a (theoretically) conservative philosopher:

> Quine's arguments in *Word and Object* show that [. . .] [r]evisability is a property only of particular sentences, or restricted sets of sentences, and [that] it cannot extend to the total system itself. [. . .] No revision open to us can take us beyond the language we now use and understand—any 'alternative' is either something we already understand and can make sense of, or it is no alternative at all. [. . .] If revisions can only be partial at best [. . .] then clearly we must rely on the on-going conceptual scheme we now possess. [. . .] And this in turn is behind Quine's apparently conservative attitude towards theoretical change and revision. [. . .] This "taste for old things" is not just an idiosyncrasy of Quine's; it is the only course open to anyone in his attempts to fit theories to experience.[21] (Stroud, 1968, pp. 92–93).

Finally, some commentators argue that at "the heart of Quine's philosophical position" lies his "attack on standard philosophical views about meaning" (Harman, 1967, p. 124); some read Quine as a "contextualist" (Thompson, 1964, pp. 211–29); some believe that Quine's "philosophical contributions are narrowly restricted to the technical field of modern logic" (Reck, 1968, pp. vi, 147); and some read *Word and Object* primarily as a work in metaphysics, construing a divide between the antimetaphysical positivists on the one hand and new metaphysicians like Quine, Strawson, Smart, and Sellars on the other (Coburn, 1964; Margolis, 1969; Rorty, 1970; Loux, 1972).

In the 1960s, in other words, there was no widely accepted interpretation of Quine's main message. Of course, many of the above readings are compatible to a certain extent: *Word and Object* can be read both as defending a version of pragmatism and as a work that attacks traditional views about meaning, both as a conservative work and as a defense of linguistic philosophy.[22] Still, this wide range of perspectives on *Word and Object* seems to be a symptom of the widespread feeling that the book does not provide "a punctual summary of [Quine's]

[21] See also Rescher (1960, pp. 376A–377A) and Grandy (1973, p. 99).

[22] Not all such interpretations are compatible, however. Where Coburn reads *Word and Object* as a "full-scale defense of realism" (1964, p. 205), for example, Grandy argues that Quine accepts realism only grudgingly because "entities are posited only when there is no way of explaining the data without such posits" (1973, p. 99). Similarly, the view that Quine's ultimate concern is "to decide what there really is" (WO, 1960, 270), is incompatible with the interpretation that his section on semantic ascent functions as a defense of linguistic philosophy (Rorty, 1967b, p. 11).

manifold thoughts" (Becker, 1961, p. 238, my translation), or, more negatively, that it is "hard to see a really coherent position behind Quine's book" (Stenius, 1968, p. 27).

Perhaps as a result of these developments, Quine was dissatisfied with the responses to his work. Not only was he displeased with the reviews of his book,[23] he was also clearly annoyed by the many misinterpretations of his central theses. In response to an author of one of the many letters and papers about *Word and Object* he received, Quine writes:

> Clear exposition, much though I admire it, is a knack that has eluded my best efforts early and late. *Word and Object* was the hardest; I struggled for years in the writing of it, trying, not primarily to clinch proofs even, but to convey something of my basic thoughts. Your paper is as devastating evidence as I have seen of the failure of that effort.[24] (September 18, 1967, item 675)

Arguably, what was missing was a catch phrase or a distinctive "ism" to summarize his core philosophical perspective. Where positivists, pragmatists, realists, functionalists, and materialists benefited from a clearly identifiable label (no matter how vaguely defined these labels typically were), Quine's philosophy was not identified with any such "ism."[25] Although Quine was well known for a wide range of philosophical theses and arguments in the early 1960s—his rejection of the analytic-synthetic distinction, his criterion of ontological commitment, his views about meaning, the Duhem-Quine thesis, his indeterminacy arguments, and his perspective on modal logic and propositional attitudes—these arguments and theses could not be easily tied together with a convenient label.[26] Indeed, in the first half of his philosophical career, Quine did not like to identify his view with a particular "ism." Already in a 1938 letter

[23] See Quine's letter to Lynwood Bryant, the director of the MIT Press: "Thanks for sending me Johnstone's review. It is sad to see one of the substantial journals come through so inadequately. Strangely, the review in *Philosophical Books* by Peter Geach was scarcely more competent. The only creditable review I've seen thus far is the sixteen-page one [. . .] by G. F. Presley" (November 16, 1961, item 1489).

[24] In response to a paper by Richard Schuldenfrei, Quine writes: "You have grasped my point of view very successfully. This is something that few readers seem to have done" (June 16, 1971, item 958).

[25] Indeed, C. F. Presley's three-thousand-word encyclopedia entry on Quine in the 1967 edition of the *Encyclopedia of Philosophy* does not ascribe any "ism" to Quine (Presley, 1967).

[26] If anything, Quine was viewed as a slayer of "isms." Antony Quinton, in an 1967 article in *The New York Review of Books* (title: "The Importance of Quine"), argues that Quine has created "a coherent theory of knowledge of great boldness and originality" by combining the logical expertise and the science-mindedness of his Viennese predecessors, while rejecting their three main "isms": phenomenalism, conventionalism, and verificationism (Quinton, 1967).

to Wilfrid Sellars, Quine quipped that his "ratio of questions to answers" was "too high at present to let me identify myself with any 'ism'" (January 18, 1938, item 972). In a note to Putnam, Quine writes: "The main fault of ismism is that it generates straw men whose isms could never have been embraced by flesh and blood" (undated, item 2388a, my transcription). The one time Quine did adopt an "ism"—i.e. in his 1947 joint paper "Steps toward a Constructive Nominalism"—he was displeased with the consequences. During much of the 1950s and 1960s, Quine had to correct the widespread impression that he was a staunch nominalist. See, for example, *Word and Object* (1960, 243 n. 5):

> A[n] [. . .] accountable misapprehension is that I am a nominalist. I must correct it; my best efforts to write clearly about reference, referential position, and ontic commitment will fail of communication to readers who [. . .] endeavor in all good will to reconcile my words with a supposed nominalist doctrine. In all books and most papers I have appealed to classes and recognized them as abstract objects. I have indeed inveighed against making and imputing platonistic assumptions gratuitously, but equally against obscuring them. Where I have speculated on what can be got from a nominalistic basis, I have stressed the difficulties and limitations. True, my 1947 paper with Goodman opened on a nominalist declaration; readers cannot be blamed.

7.5. Adopting an "Ism"

In the mid-1960s, however, Quine's attitude changed, perhaps as a result of the mixed reviews and the confusion surrounding the overarching perspective offered in *Word and Object*.[27] Where Quine had never used the term "naturalism" or the slogan "no first philosophy" before 1965, these phrases become omnipresent in his work from the late 1960s onward. Between March and July 1968 alone, for example, Quine uses these phrases to describe his perspective in no fewer than seven lectures, essays and responses:

- "Natural Kinds," first draft completed March 6, 1968 (item 1355): "my position is a naturalistic one; I see philosophy not as an a priori propaedeutic or

[27] Even Quinton (1967), in his very sympathetic assessment of Quine's philosophy in the *New York Review of Books* (see footnote 26), argues that "*Word and Object* was something of a disappointment," as Quine's general philosophical view about knowledge "needs and deserves a fuller and more systematic exposition."

groundwork for science, but as continuous with science. [. . .] There is no external vantage point, no first philosophy."

- "Ontological Relativity," presented at Columbia University March 26–28, 1968 (item 2995): "Philosophically I am bound to Dewey by the naturalism that dominated his last three decades. [. . .] There is no place for a prior philosophy."
- "Epistemology Naturalized," first draft dated "March 1968" (item 2441): "The old tendency was due to the drive to base science on something firmer and prior in the subject's experience; but we dropped that project [dislodging] epistemology from its old status of first philosophy."
- Reply to Smart, completed June 12, 1968 (item 1490): "The key consideration is rejection of the ideal of a first philosophy, somehow prior to science."
- Reply to Chomsky, completed June 12, 1968 (item 1490): "There is no legitimate first philosophy, higher or firmer than physics, to which to appeal over physicists' heads."
- Symposium with Sellars, July 11, 1968 (item 2903): "We are both naturalists. We see philosophy as continuous with science and hence as a going concern in the real world of bodies."
- Reply to Stenius, completed July 30, 1968 (item 1490): "I am able to take this stance because of my naturalism, my repudiation of any first philosophy logically prior to science."[28]

Interestingly, earlier versions of both "Epistemology Naturalized"—at the time still called "Stimulus and Meaning" (November 1965, item 2756) and "Empiricism in Transition" (January 1967, item 3102)—and "Ontological Relativity" (March 1967, item 2994) do *not* yet contain the term "naturalism."

Sometime in the mid-1960s, in other words, Quine decided to label his philosophy "naturalistic."[29] Arguably, Quine's decision was influenced by

[28] These essays and responses (except Quine's contribution to the symposium with Sellars) were later published as (NK, 1969c, 127), (OR, 1968a, 26), (EN, 1969b, 87), (RSM, 1968d, 265), (RC, 1968f, 275), and (RST, 1968e, 270) respectively. Quine's adoption of the label "naturalism" proved to be very effective. From the early 1970s onward, Quine's philosophy starts to be widely described as naturalistic. See, for example, Kmita (1971), Giedymin (1972), Wilder (1973), Hockney (1973), and Greenlee (1973). Before 1968, I have found no description of Quine's view as naturalistic. The only exception is Thompson (1964, 213), who writes that Quine "appears close to a form of contextualistic naturalism."

[29] It is difficult to establish when Quine exactly adopted the label "naturalism" for his own philosophy, but he does describe his view as "naturalistic" in a September 22, 1967, talk entitled "Kinds" (item 2948). The first use of the term "naturalism" in its contemporary sense I have been able to trace is from June 29, 1966, when Quine, in a draft of "Russell's Ontological Development" characterizes the evolution of *Russell's* views by claiming that "there is an increasing naturalism, an increasing readiness to see philosophy as natural science trained upon itself and permitted free use of scientific findings" (item 2733). Before 1966, Quine does speak about "naturalistic arguments"

the many misinterpretations of his position. The most influential series of misreadings was due to Noam Chomsky, who dismisses Quine's views in his paper "Quine's Empirical Assumptions" (1968). Quine, who must have immediately remembered the impact of Chomsky's review of Skinner's *Verbal Behavior* (1959), was clearly alarmed; within seven months after he first received Chomsky's paper on November 16, 1967 (item 1490), Quine writes and presents three papers in which he responds to these arguments: "Philosophical Progress in Language Theory" (first delivered in February 1968), "Linguistics and Philosophy" (presented on April 13, 1968), and "Reply to Chomsky" (completed on June 12, 1968).[30]

"Naturalism" was not the only "ism" Quine considered in the late 1960s. In a 1965 note entitled "The Sophisticated Irrational," Quine worries about the tendency among recent philosophers (he mentions N. R. Hanson, the gestalt psychologists, and Thomas Kuhn) to get "carried away" in their opposition to positivism. Quine notes that although "certain hopeful positivistic tenets turned hopeless," the positivists also provided "[v]aluable supervening insights." According to Quine, people like Hanson and Kuhn "forget what they came for" like an "invading army destroying conquered riches." In response, as we have seen in chapter 4, Quine offers what he calls "*involutionism*" or "*immanentialism*," the idea that "[w]e must work within a growing system to which we are born." According to Quine, Hanson and the gestalt psychologists fail "to see that involutionism & behaviorism take care of obsvn. sentences to perfection," whereas Kuhn's "cultural relativism," the idea that there is "no truth," is also solved by involutionism via Tarski's theory of truth (July 21, 1965, item 3182, my transcription).[31]

Apart from "involutionism" and "immanentialism," Quine also toyed with the labels "psychologism" and "externalized empiricism." Indeed, one draft of "Epistemology Naturalized" is titled "Epistemology Naturalized; or, the Case for Psychologism" (item 2441), and Quine talks about "externalized empiricism" in "Linguistics and Philosophy" (LP, 1968i, 58), in "Philosophical Progress in Language Theory" (PPLT, 1970b, 7), and in a draft for a lecture at Rockefeller University, where Quine spent a term in 1968: "[The] change of viewpoint that I have been describing is not an abandonment of empiricism, as I would use the term, but it is an externalizing empiricism. Instead of taking subjective sense

and "naturalists" in (LDHP, 1946c, 112) and (MSLT, 1953b, 149) respectively, but his use of those terms there is unrelated to the metaphilosophical view with which he identifies the term in the later stages of his career.

[30] See (PPLT, 1970), (LP, 1968i), and (RC, 1968f).

[31] A transcription of "The Sophisticated Irrational" is included in the appendix (A.6).

impressions and trying to reduce all talk of external objects to that basis, we start with the objects" (February 5, 1968, item 2902).[32]

So why did Quine, in the end, choose the label "naturalism" over "immanentialism," "psychologism," "involutionism," and "externalized empiricism"? It is hard to give a conclusive answer to this question, but there is some evidence that the answer lies in his preparation of "Ontological Relativity" for his John Dewey Lectures in March 1968. When Quine first presented "Ontological Relativity" at Chicago and Yale in May 1967, his paper, as noted above, did not yet contain the term "naturalism." Nor did the paper contain any reference to Dewey. In the invitation for the John Dewey Lectures, however, Quine was explicitly asked to deal "with some aspect of John Dewey's work" or to apply "the principles or the spirit of his philosophy to some field of human thought or endeavor" (January 10, 1967, item 248). When Quine adapted his paper on ontological relativity to suit the lecture series, therefore, he needed a reference to Dewey. Now, it is generally known that Quine was not well versed in the work of John Dewey.[33] Given that the paper Quine eventually read at Columbia in March 1968 (item 2995) refers only to Dewey's *Experience and Nature*, it is likely that he read only this work in order to draw a connection, a work that literally opens with the statement that "[t]he title of this volume [. . .] is intended to signify that the philosophy here presented may be termed either empirical naturalism or naturalistic empiricism" (Dewey, 1925, p. 1). Indeed, Quine's book collection (stored at Houghton Library) contains only one of Dewey's books—*Experience and Nature*. In the back of this copy, Quine has summarized the first pages of the book with the note: "naturalism as opposed to trad'l epistemology" (undated, AC95.Qu441.Zz929d, my transcription).

[32] One might wonder at this point whether Quine only adopted a new label for his philosophy or whether the new description of his views represents a substantive change of mind. Quine himself has claimed that there was no change of mind. In response to a letter from Shapour Etemad, who reported that Chomsky had argued that "Quine adopted [. . .] naturalized epistemology" in response to "the critique of his position in *Word and Object*," Quine writes that "[t]he statement that you report from Chomsky is puzzling. *Word and Object* was already utterly naturalistic. 'Epistemology Naturalized; [. . .] differed only in explicitly proclaiming naturalism as a policy in epistemology" (August 1989, item 336, my transcription). Quine's recollection is confirmed by his own 1968 description of *Ontological Relativity and Other Essays* to the marketing department of Columbia University Press: "Philosophical doctrines propounded in my *Word and Object* and earlier writing aroused considerable controversy, some of which is traceable to misunderstanding. The six essays assembled in this book are meant in part to resolve misunderstanding and in part to press the doctrines further" (October 1968, item 1422, my transcription).

[33] See Koskinen and Pihlström (2006) and Godfrey-Smith (2014). Indeed, a few months after his John Dewey Lectures, Quine admitted that he was "not much of a Dewey scholar" in a letter to Pearl L. Weber (September 13, 1968, item DVFM 43/16208, Hickman 2008). See also Quine (IWVQ, 1994b, 70): "I was not influenced by Dewey. I didn't know his work that well in the old days."

A second possible influence on Quine's decision to describe himself as a "naturalist" in the John Dewey Lectures is Ernest Nagel, the first John Dewey Professor in Philosophy at Columbia. At the time, Nagel was well known as a "naturalist." Nagel had defended naturalism in his 1954 presidential address to the APA Eastern Division Meeting (title "Naturalism Reconsidered") and he had coauthored a defense of naturalism with John Dewey (and Sidney Hook) in the 1945 article "Are Naturalists Materialists?"[34] When Quine received an invitation to be the first Dewey lecturer in January 1967, Nagel and Quine had already been close colleagues for thirty years. Nagel had tried to find a position for Quine in 1939 when the Harvard administration turned down the department's request for his promotion (May 1939, item 758), Nagel had asked Quine to replace him when he was on leave during the 1950–1951 academic year (December 20, 1949, item 248), and Quine had urged Donald C. Williams that they "should do all we can to get Nagel" to Harvard in 1954, arguing that he would "obviously [. . .] be a tremendous acquisition" (April 9, 1954, item 471). Perhaps Quine's decision to adopt the label "naturalism" in his Columbia lectures was also influenced by the fact that Nagel was at Columbia.

Of course, Dewey's and Nagel's "naturalisms" differ somewhat from the way in which Quine uses the term. Whereas "Columbia naturalism"[35] primarily opposes dualism and supernaturalism, Quine's rejection of first philosophy is predominantly a response to the view that one can strictly distinguish science and philosophy. It is questionable, however, whether Quine was already sensitive to this distinction in the late 1960s. Indeed, he seems to conflate metaphysical naturalism, methodological naturalism, and metaphilosophical naturalism in the passage in which he connects his views to Dewey's:

> Philosophically I am bound to Dewey by the naturalism that dominated his last three decades. With Dewey I hold that knowledge, mind, and meaning are part of the same world that they have to do with, and that they are to be studied in the same empirical spirit that animates natural science. There is no place for a prior philosophy. (OR, 1968a, 26)

This passage still contains a metaphysical thesis ("knowledge, mind, and meaning are part of the same world that they have to do with"), a methodological thesis (knowledge, mind, and meaning are to be studied "in the same empirical spirit that animates natural science"), and a metaphilosophical thesis ("there is no

[34] See also Danto's entry on naturalism and Thayer's entry on Nagel in the 1967 edition of the *Encyclopedia of Philosophy*.

[35] See Eldridge (2004) and Jewett (2011). The term was already in use in the late 1960s. See, for example, Reck (1968, ch. 4).

place for a prior philosophy").[36] In later work, however, metaphilosophical naturalism takes center stage in Quine's philosophy and is argued for on independent grounds: there is no distinction between science and philosophy because "we must work within a growing system to which we are born" and because there are no transcendental, distinctively philosophical perspectives from which to question this system.

Whatever the exact reasons behind Quine's adoption of the term "naturalism," however, he clearly decided to stick with the label. When he was invited to give one of the principal addresses at the World Congress of Philosophy in 1968, he seems to have decided to use the stage to definitively rebrand his philosophy as "naturalistic." He rewrote his lecture "Stimulus and Meaning" and renamed it "Epistemology Naturalized." Before he read his lecture, Quine addressed his Viennese audience in German, marking the occasion as follows:

> Es freut mich sehr, zum ersten Mal seit 35 Jahren nach Wien zurückzukommen. Im Jahre 1932 bekam ich in Amerika mein Doktorat und zugleich ein Reisestipendium, welches mich nach Wien brachte. Damals waren Schlick und Waismann hier und auch Gödel, der eben seine welterschütternde Entdeckung veröffentlicht hatte. Carnap war eben nach Prag übergesiedelt, wohin ich ihm fünf Monate später folgte. Seine *Logischer Aufbau* war in den Läden und seine *Logische Syntax* in der Schreibmaschine. Ich wurde ein ernster Jünger Carnaps. Nun sind 35 Jahre vorbeigegangen. Meine philosophische Stellung ist beträchtlich geändert worden und so ist auch die Carnapsche, obwohl nich in der gleichen Richtung. Gewisse Unterschiede zwischen meinem heutigen Gesichtspunkt und der altwienerischen Philosophie werde ich jetzt betrachten.[37] (September 9, 1968, item 1239)

[36] In a note he wrote while preparing "Ontological Relativity," Quine does seem aware that there are different *arguments* for naturalism. He distinguishes between a negative argument for naturalism, which hinges on the "failure of phenomenalism" (i.e. there is no distinctively epistemological perspective on reality) and a positive argument which "hinges on meaning & the nature of language" (i.e. language, and hence meaning, is a natural phenomenon) (ca. September 1967, item 3182, my transcription). See also Quine's response to Grover Maxwell (1968c), written in March 1966 (item 616), in which he uses the term "naturalistic" to describe the view that language and mind are part of the natural world (RGM, 1968c, 176).

[37] "I am very pleased to return to Vienna for the first time in 35 years. In 1932, I received my doctorate as well as a travel scholarship, which took me to Vienna. At the time, Schlick and Waismann were here, as well as Gödel, who had just published his world-shattering discovery. Carnap had just moved to Prague, where I followed him five months later. His *Logical Structure* was in the stores and his *Logical Syntax* in the typewriter. I became a serious disciple. Now, 35 years have passed. My philosophical position has changed considerably, and so has Carnap's, though not in the same direction. I will now reflect on the differences between my present point of view and the Old Viennese philosophy" (my translation).

Although "Epistemology Naturalized" failed to expose Quine's subtle views on the relation between science and philosophy as clearly as one might wish (see chapter 2), Quine's new label had tremendous impact on the course of analytic philosophy. Among present-day philosophers, metaphilosophical naturalism is widely viewed as a predominantly Quinean tradition.

7.6. Conclusion

In this chapter I have examined the development and the reception of Quine's naturalism after he had rebooted his book project 1952. I have argued that internal tensions and external criticisms forced Quine to continuously develop, rebrand, and refine his metaphilosophy before he eventually settled on the position that would spark the naturalistic turn in analytic philosophy. Although Quine had already concluded that epistemology should be viewed as an empirical science in 1952, he started to turn to the sciences only in the late 1950s. Finally, I have shown that the reception of *Word and Object* outside Quine's closest philosophical circles was confused, and I have argued that Quine, perhaps in response to these misunderstandings, decided to adopt the label "naturalism" for his philosophy in the late 1960s.

8

Conclusion

The history of twentieth-century analytic philosophy is often depicted as a history of two broadly diverging projects: the formal, science-minded philosophy of Russell and the logical positivists and the common-sense philosophy of Moore and the ordinary language philosophers. Quine's philosophy, though distinct from the positivists in its rejection of the two dogmas, in many respects falls into the former category. Not only are Quine's views undeniably science-minded, he also adopts the positivists' rigorous methodology, i.e. their insistence on employing the standards of clarity and precision that are most commonly displayed in the natural and the formal sciences.

Given this coarse-grained classification as well as his strong emphasis on the continuity of science and philosophy, it is not surprising that Quinean naturalism is often depicted as some sort of scientism in sheep's clothing—as an implausibly deferential worldview in which every type of inquiry is ultimately evaluated in terms of its merit for the sciences narrowly conceived. Indeed, even contemporary naturalists often equate Quinean naturalism with a staunch rejection of normative epistemology and an austerely eliminativist form of physicalism.

In this book, I have argued that this is a mistake. I have argued that at the heart of Quine's naturalism lies the adoption of a radically *immanent* perspective. To be a Quinean naturalist is to work from within. Or, as Quine himself describes it in one of his last philosophical papers, to be a Quinean naturalist is to live "within one's means" (NLOM, 1995c). Quine dismisses some traditional philosophical questions because they presuppose an *external* perspective, not because they cannot be reduced to natural science.

Reading Quine's naturalism as a radically immanent philosophy sheds new light on his arguments against traditional epistemology and metaphysics. Quine does not reject the traditionalists' views out of despair but because they crucially presuppose an extrascientific vantage point. In epistemology, both the skeptic who questions our knowledge about the external world and the empiricist who tries to justify this knowledge by reconstructing it from sense data rely on the viability of such a transcendental perspective. The skeptic presupposes that she can challenge science without presupposing science, whereas the empiricist presupposes that she can answer this challenge by reducing our theory to some science-independent sensory language. In a similar fashion, Quine dismisses traditional metaphysics. For in asking what reality is really like, the traditional metaphysician is asking us to set aside our ordinary conceptions of truth and reality.

In response to the question as to wherein our perspective should be immanent, Quine's answer, of course, is "science." It is partly because of this reason that his naturalism is often interpreted as scientistic. I have argued, however, that such a reading presents Quine's philosophy the wrong way around. Even if we, like Quine, adopt a strictly science-immanent perspective, a question that remains to be answered is how broad our conception of science ought to be. Quine's notion of science, we saw, is extremely broad, as it encompasses our complete "inherited world theory," including common sense. Quine's naturalism, in other words, only presupposes that we start in the middle. It is only *after* adopting such a broadly science-immanent perspective that Quine, in regimenting his system, starts making choices that some contemporary philosophers have argued to be unduly restrictive. Whether or not we end up with a scientistic philosophy, in sum, depends on the choices we make *while* trying to "improve, clarify, and understand the system from within" (FME, 1975a, 72).

My interpretation of Quine's naturalism is supported by the way in which he developed this position. In the second part of this book I have shown how Quine grew from a respected logician to a world-leading philosopher. Reconstructing the way in which Quine *himself* improved, clarified, and understood his system from within, I have argued that there is no one event that changed his philosophical outlook. Rather, Quine's development was gradual; numerous events played a key role in maturing his perspective: (*a*) his adoption of the criterion of ontological commitment in 1939; (*b*) the evolution of his views on metaphysics in the early 1940s; (*c*) his renewed view about his disagreement with Carnap in 1943; (*d*) his adoption of a wide-scoped holism in the late 1940s; (*e*) his epistemological breakthrough in 1952; (*f*) his development of a genetic project in the mid-1950s; (*g*) his increased concern with psychology and linguistics in his preparation of *Word and Object*; (*h*) his growing realization that science and philosophy are continuous in the 1950s; and finally (*i*) his decision to adopt the label "naturalism" in the late 1960s. Although most of these developments can be explained in terms of the difficulties Quine encountered while working from within the radically empiricist view he had advanced since his student years, they all gradually pushed him toward the position that there are no transcendental perspectives on our inherited world theory whatsoever—toward the position that there is no "view from somewhere else."

When it comes to his naturalism, therefore, Quine's position is more closely aligned with common-sense philosophy than has often been supposed. Even though Quine and the ordinary language philosophers exhibit a difference in style as well as a difference in emphasis—in *Word and Object*, Quine worries that ordinary language philosophers "exalt ordinary language to the exclusion of one of its own traits: its disposition to keep on evolving" (WO, 1960, 3)—both acknowledge that we ought to start in the middle and eschew transcendental perspectives (p. 4). Indeed, in reflecting on the role of naturalism in twentieth-century analytic philosophy, Quine claims that "certainly the ordinary language philosophers [. . .] are as naturalistic as one could wish" (1993, item 2498).

Something similar can be concluded about the relation between Quine and the logical positivists. Whereas the Carnap-Quine debate has often been depicted as involving a fundamental disagreement about the nature of inquiry, there is every reason to suppose that they share the view that, as philosophers, we are always working from within. Indeed, I have argued that Quine never viewed Carnap as a first philosopher who aims to validate our scientific theories from an external perspective. Carnap, too, dismissed the "Cartesian dream" and transcendental notions of "truth" and "reality."

Still, Quine's position should not be viewed as a mere synthesis of logical positivism and common-sense philosophy. While both branches of analytic philosophy implicitly seem to accept the view that one ought to "live within one's means," they maintain a strict distinction between science and philosophy—between matters of fact and matters of language. Quine's contribution, therefore, lies not so much in the *nature* of his approach as in the rigor with which he extinguishes transcendental philosophizing. Quine's adoption of an absolute science-immanent approach led him to dismiss a broad range of widely accepted philosophical presuppositions and distinctions. To name only a few of the examples I have discussed: he adopted a strictly deflationary theory of truth, an austerely minimalist conception of justification, and a new view about the nature of artificial languages, and he rejected the distinctions between the analytic and the synthetic, between the meaningful and the meaningless, between empirical and mathematical knowledge, between internal and external questions, and between science and philosophy. Quine's naturalism, in sum, is radical not because it is immanent to *science*, but because it is science-*immanent* across the board.

APPENDIX

A.1. Editorial Introduction

In this appendix, I have collected and transcribed the archival documents that have most affected my views about Quine's naturalism and its development.

The first document, "Logic, Mathematics, Science" (December 20, 1940, item 2954) is Quine's lecture for the 1940–1941 Harvard Logic Group. The lecture was read in the presence of Carnap, C. I. Lewis, Donald C. Williams, Paul Marhenke, Nelson Goodman, Albert Bennett, Charles Baylis, and Clifford Barrett on December 20, 1940. Carnap's notes of the Harvard Logic Group meetings between March 1940 and June 1941 have been published and introduced in Frost-Arnold (2013). The present document illuminates Quine's contributions to the Logic Group discussions, as the lecture sets out his views about the finitist language the group tried to develop. Furthermore, the lecture sheds light on Quine's views about the relation between logic, science, and common sense in the early 1940s (see chapters 5 and 6).

"Sign and Object; or, The Semantics of Being" (October and November 1944, items 3169 and 3181) is a set of four autograph notes written between October 4 and November 5, 1944, when Quine was a lieutenant in the navy. As I explain in chapter 5, Quine worked on a philosophical monograph with the title *Sign and Object* between 1941 and 1946. This set of notes contains Quine's most detailed discussion of metaphysics, epistemology, and the task of philosophy from this period. Furthermore, the note is remarkable because it contains some passages that appear verbatim in *Word and Object*, published almost sixteen years later.

Morton White's *A Philosopher's Story* (1999) contains a triangular correspondence between Quine, White, and Nelson Goodman from 1947. This series of letters has often been used to analyze Quine's views about analyticity in the late 1940s. The third item, "An Extensionalist Definition of Meaning," (January 1949, Papers of Nelson Goodman, Harvard University Archives, accession 14359) contains an exchange of two letters between Quine and Goodman from January 1949 which reveals their attitudes *after* this triangular correspondence. The letters show that Quine had not definitively abandoned his attempts to find a behavioristically acceptable definition of analyticity in the year before he wrote "Two Dogmas of Empiricism." The letters play an important role in my reconstruction of Quine's evolving ideas about analyticity in the 1940s (chapter 6).

"The Present State of Empiricism" (May 26, 1951, item 3015) is an autograph paper Quine read at the American Academy of Arts and Sciences in Boston on May 26, 1951, at an event organized to discuss "Two Dogmas of Empiricism," the paper Quine had published a few months earlier. In his lecture, Quine reflects on the nature of his achievement in "Two Dogmas"; surprisingly, he argues that the paper does not contribute "a new idea to philosophy" (see section 5.8). Furthermore, Quine argues that the doctrine of reductionism is "the most basic of the two" dogmas (see section 6.1). After Quine's lecture, "Two Dogmas" was discussed by Gustav Bergmann and the panelists Percy Bridgman, Herbert Feigl, Henry Margenau, Richard von Mises, Hao Wang, Yehoshua Bar-Hillel, Philipp Frank, Ernest Nagel, and John C. Cooley. Quine's autograph notes of his responses to these commentators are also added to the transcription. Some of these notes are discussed in chapters 5 and 6.

"The Sophisticated Irrational," (July 21, 1965, item 3182) finally, is a short autograph note from July 21, 1965. In the note, Quine argues that some critics of logical positivism—notably N. R. Hanson and Thomas Kuhn—are throwing away the baby with the bathwater. Furthermore, Quine introduces the term "involutionism" to describe the view that "[w]e must work within a growing system to which we are born" (see chapter 4). In chapter 7, the note is used to illustrate Quine's attempts to rebrand his philosophy in the mid-1960s.

A.1.1. Editorial Remarks

In transcribing Quine's autograph notes and drafts, I have aimed to minimize editorial interference and chosen not to correct ungrammatical shorthand, except in transcribing the documents' titles.[1] I have not transcribed words, sentences, and paragraphs that are crossed out. In transcribing "Logic, Mathematics, Science" and the letters in "An Extensionalist Definition of Meaning," the

[1] Quine's 1940 lecture as transcribed is titled "Logic, Math., Science" in the original document. The title of the third selection ("An Extensionalist Definition of Meaning") is mine.

only typed documents in the series, I have incorporated Quine's autograph corrections as well.

In some of the manuscripts, Quine uses both parentheses () and square brackets []. All editorial remarks are between braces {}. The asterisked footnotes are Quine's; the footnotes listed by numerals are mine. Books and articles mentioned by Quine are included in the bibliography of this book.

A.2. Logic, Mathematics, Science (1940)

READ BEFORE CARNAP, LEWIS, WILLIAMS, MARHENKE, GOODMAN, BENNETT, BAYLIS, AND BARRETT AT I.A. RICHARDS' OFFICE, DEC. 20, 1940

In order to develop any considerable theory in the way of a general semantics of formalized languages, it is necessary to impose some limitations or other on the variety of language patterns studied. A *completely* general semantics is surely doomed to triviality.

Anyway, we shouldn't mind limiting our object languages in a fashion that excludes only trivial or uninteresting languages. Again we shouldn't mind a limitation which excludes certain interesting, non-trivial languages, so long as it continues to admit other languages into which those excluded ones are translatable. Our study of the excluded forms can still proceed indirectly, through study of their more canonical translations.

Last time I recommended, in company with Tarski, that this course be carried to an extreme: that we focus our attention on those languages which involve just constant predicates, joint denial, and universal quantification with respect to a single syntactical category of variables.

The formulae of such a language comprise atomic formulae—consisting each of an atomic predicate followed by variables—and in addition they comprise everything that can be built up of atomic formulae by joint denial and quantification. Such a formula I shall call *normal*: and normal formulae without free variables I shall call normal statements.

The question is, of course, whether this is too much of a restriction. I have several brief comments to make in this regard.

In particular, all math. is translatable into such a language. Here the sole predicate is the two-place membership predicate "з."

So is syntax. For protosyntax we take just the three-place predicate "M," with help of which we can define concatenation and identity and names of individual signs. For broader syntax we throw in "з."

The lack of constant terms and functors is no hindrance, for these can always be got with help of appropriate predicates by contextual definition.

Two advantages of so doing were pointed out in §27 of *M.L.*:[2] (1) a technical advantage, viz. quantification theory can be simplified and can be kept standard for all applications so long as we admit no terms but variables, and (2) a philosophical advantage, viz. questions of meaningfulness become wholly dissociated from questions of existence.

The question whether *all* languages ever needed are so translatable turns on the so-called thesis of extensionality: the thesis that no one formula (statement or matrix) ever needs occur in another save as argument to a truth function or scope of a quantifier. The only current formalized languages resisting such translation are those involving modalities.

And there is the converse question: Are there systems, worthy of serious study, which are so weak that on translation they fail to exhaust any language of the described kind? In other words, systems not even containing truth functions and quantification, or translations thereof? I doubt that any such need consideration. Even elementary arithmetic, or protosyntax, or Huntington's various simple systems (betweenness, also so-called math'l theory of strict implication, etc.), or Euclid's *Elements*, are strong enough. Systems resisting such regimentation can probably be studied well enough in isolation if and when occasion arises, anyhow; on the other hand the known utility of metatheory seems to me to be preserved under the confinement of object language which I[3] propose, and of course semantics in this form can be developed more simply and in fuller detail when thus focused.

Now I propose, provisionally, to limit the term "logic" to the theory of joint denial and quantification, and to classify "з" rather as the basic non-logical device of classical mathematics (cf. Carnap, *Abriß*, §31).[4] What I am calling logic could be called elementary logic, if the broader sense of "logic" seems too well established to be abandoned; still, I think my usage is more in the spirit of the *long* tradition than the broader sense is. This is a verbal question, but my purpose is more than verbal: I want to put special emphasis on the boundary which separates logic in this narrow sense from the rest of science. This will help clarify the general attitude which I want to express.

This notion of logic is, more accurately, a notion of logical truth: a normal statement is logically true when it remains true under all uniform changes of its predicates. This definition involves a difficulty, actually, in that the totality of predicates is relative to one or another extra-logical language. If "is brother of"

[2] See (ML, 1940). *Mathematical Logic* (publication date: September 5, 1940, item 1391) had just been published when Quine presented "Logic, Math., Science" to the Harvard Logic Group.

[3] Editorial correction of "WhichI."

[4] Carnap (1929).

were the only predicate, then the non-transitivity principle would become logically true, which we don't want. But this is easily fixed; I'll return to it.

Now I want to mention some of the important differences between logic, in the present sense, and other matters.

There are no logical statements, for there are no logical predicates. Studies of logic proceed metatheoretically, as in first 2 chh. of *M.L.* I don't object to "p," "q," etc., and "F," "G," etc., but these are not variables in the pronominal or bindable sense; expressions containing such letters are not formulae—neither statements nor parts of statements—but in my opinion are best regarded merely as diagrams auxiliary to discussions in metatheory. In effect, whether by these devices or by more directly metamathematical notation, we can say only that all statements of such and such form are logically true or logically imply such and such others; we can't give examples without bringing in predicates from outside of logic.

As soon as even "3" comes in, the case is otherwise: we can construct actual statements. We get only then, we might say, a content—a mathematical subject matter—whereas logic itself touches only the outward forms of the statements.

Another importance of the boundary: logic calls for no objects in particular. Ontological demands enter only with predicates. This is not because predicates designate; predicates themselves remain syncategorematic and devoid of designations in any case, being never replaceable by variables. *But* adoption of a predicate may make ontological demands because of the objects for which the predicate calls as values of the attached variables—in short, the objects which *fulfill* the predicate. The predicate "3" makes ontological demands, calling as it does for classes—universals. Thus it is that mathematics is Platonic, while logic is not.

Logic doesn't even depend on the size of the ontology. The normal statements which would be logically true from the standpoint of an infinite universe would be logically true from the standpoint also of all finite universes and vice versa.

The logical truths would seem to be true for nearly all philosophies—nearly all points of view—including nominalism and realism alike. Intuitionism and related views would seem to constitute the sole exception; and even these attitudes could perhaps be satisfied by extra-logical restrictions, viz. by extruding predicates of non-constructionalistic sorts in some sense or other of the term).

Logical truth can be defined syntactically, indeed protosyntactically, in view of Gödel's completeness proof. We may just specify an infinite set of so-called axioms of quantification, as in *M.L.*[5] (for any choice of predicates), and then describe the logical truths as comprising these and all statements obtainable from these by one or more steps of modus ponens. Besides being more elementary

[5] Emphasis added.

than the semantic characterization in terms of truth, this is free from the difficulty noted in that connection.

This gives a precise and yet general notion of logical truth, insofar as my originally proposed regimentation of the object languages of semantics can be maintained. Of course it carries with it no epistemology of logical truth, such as conventionalism or intuitionism or empiricism; I am as ignorant as ever of the meanings of any of these seemingly mutually opposed theories of the epistemology of logic.

Logical consequence is definable in terms of logical truth: S_1 logically implies S_2, or S_2 is a log. consequence of S_1, when the conditional having S_1 as antecedent and S_2 as consequent is logically true. Or, equivalently, when S_2 is derivable by iterated modus ponens from S_1 and the axiom of quantification.

[Seminar: insert pp. 5a–6][6]

We get one or another extra-logical notation (math'l, biological, etc.) by introducing appropriate predicates. We get elementary arithmetic, e.g., by introducing "P" where "Pxyz" means that $x = y^z$. We can define identity easily, then description contextually, and finally y^z as $(\iota x)Pxyz$, and $x{\cdot}y$ as $(\iota z)(w)(w^z = (w^x)^y)$ and $x + y$ as $(\iota z)(w)(w^z = w^x{\cdot}w^y)$.

We get all math. by introducing "з." We may attempt to axiomatize part of it, with a finite set of axioms as with Bernays (fol'g von Neumann) or by specifying an infinite set of axioms of membership (cf. *M.L.*);[7] and our mathematical theorems are those statements which are logically implied by the mathematical axioms. But the theorems will be a fragmentary set of mathematical truths at best, as Gödel has shown.

Logic in the present sense seems to be the common denominator of all not wholly trivial theories—e.g. elementary arithmetic, protosyntax, mathematics, Huntington's betweenness theory, Euclid's geometry, etc. "з" falls outside this common ground.

Now consider a universal language for present day science—made by adopting a vast array of predicates. The ontology—the values which the variables must be regarded as taking on in order to provide for the fulfillment of the predicates in desired fashion—will include a great variety of objects:

electrons, atoms;
bacteria;
tables, chairs;

[6] Unfortunately, I have not been able to determine which pages Quine wanted to insert here.

[7] Emphasis added.

objects which may be regarded as still more immediate than tables and chairs—e.g. sense qualities such as brown or bitter, not thought of as properties of things but as objects epistemologically anterior to things;

and up at the other end there are still less real (*re*-al, thingly) objects—centimeters, distances, temperatures, electric charges, energy, lines, points, classes (or properties).

Some of these upper unthingly objects are disliked, and attempts are made to eliminate them in favor of others. E.g., Whitehead thought points, lines, and surfaces less desirable than finite volumes and classes, and hence reduced them to classes of volumes. But volumes must be interpreted, for this purpose, otherwise than as concrete composites of electrons; continuous gradation is needed. And of course classes are not concrete composites of electrons either. None of these objects are thingly, if we think of the thingly as leaving off say after concrete composites of electrons. Still, some unthingly objects seem rather less ungodly than others, and accordingly Whitehead's construction is regarded by many as a gain.

To take another example, Carnap and Jeffreys regarded distances and temperatures as less clear than pure numbers, and accordingly eliminated them contextually. But numbers, being classes, are likewise unthingly.

What I want to stress is that all universals seem to me to be on the same footing as points, centimeters, electric charges, etc.

Unthingly objects tend to be created roughly in the image of things. Lines are like wires, but infinitely thin; points are like specks, but infinitely small; energy is something like a fluid.

And classes are perhaps no exception. A class is perhaps created in the image of a heap, i.e. a thing whose surface even in the large is disconnected. "Member of a class" becomes analogous to "connected part which is not part of any other connected part." Of course classes can't be construed as heaps, any more than energy can be construed as water.

I don't insist on eliminating classes or other unthingly objects. It is not clear that the unthingly can be eliminated without losing science. Anyhow, the attempt to do so would be just an attempt to reduce the obscurer to the clearer, and to such an end there would perhaps be no reason for caring to rest with the thingly. Some things, such as electrons, would by the same token want to be defined away in favor of larger things; and as our ultimate purpose we might want to get rid of things altogether in favor of phenomena of some more immediate sort. And incidentally, the direction of clarity or epistemological priority isn't itself clearly marked out.

Rather, I think we must follow Carnap in accepting the non-positivistic or non-phenomenalistic locutions of science as uneliminable. I expect he is right in holding that we can make no more than *partial* clarifications, short of definitional eliminations. This we do by exploring relations of confirmation between statements of the more remote and statements of the more immediate sort.

Science is full of myth and hypostasis, and perhaps the purpose of its myths is in principle the same as that of earlier, more naïve myths: to imbed the chaotic behavior of ordinary things within a less chaotic and hence more intelligible super-world. The final service is prediction with regard, again, to ordinary things; but this is psychologically feasible only because of the greater Uebersichtlichkeit of the super-world which science sets up as intermediary device.

Actually the dichotomy which I have just now made between science and common sense is surely false. Common sense must no doubt be regarded as related to still more immediate experience in much the same way in which science is related to common sense; or, still more properly, there is presumably not even a trichotomy but a gradation. Tables and chairs are hypostases in the same way as are electrons, perhaps, but to a less degree.

The super-world, to revert for convenience to the rough trichotomy, gives inference for the ordinary world but not vice versa; the one is underdetermined by the other. Similarly the ordinary world is underdetermined by experience of the more or less immediate sort.

And I think of mathematics in the same way. The theory of real numbers checks with that of algebraic or of rational numbers where it touches it. The general theory of classes gives rise to ordinary common-sense results as regards finite classes—results which run parallel to common-sense laws about heaps and their maximum singly connected parts. But these latter matters do not uniquely determine the general theory of classes uniquely. Thus it is that after reading about the logical paradoxes I found myself consciously choosing a myth—Russell's myth of type theory, or Zermelo's alternative myth, or some more congenial myth of my own making. Such reflections as these, incidentally, make Gödel's incompletability theorem seem less anomalous to me than it used to.

A.3. Sign and Object; or, The Semantics of Being (1944)

A.3.1. The Reality of the External World

There is a sense in which physics might be said to be concerned with exploring the nature of reality.

And who contests this? Primarily the Idealist, claiming . . . [state grounds]. The Idealist would take the perceptions etc. rather as the basic reality, and derive things as constructions, logical constructs [Russell]. The study of how to make these constructions is Epistemology. And things are composed not of atoms but of perceptions, sense qualia, etc.

Is the Idealist then contradicting the physicist? No; and the resolution of the conflict should have been evident: things are *physically* composed of atoms, but are *epistemologically* (or: ideas of things are epistemologically) composed of qualia etc.

Two senses of real. Reason there seemed to be a rival claim for reality in an absolute sense is that an argument can be made for reducing the one to the other, and a counter-argument can be made for the opposite reduction.

But both reductions are right. Depends on purpose. And each purpose is good: realistic basis to clarify foundations of epistemology (e.g. that only light-rays etc. are given; nature of epistemological priority; etc.) and vice versa (for, as Carnap says, physicist retreats in epistemological direction when statements are challenged).

Needless, though, to use the word "reality" here. Epistemological primacy for the one; and what there is, simply—"reality," if you like—for the other. So I *am* being a realist rather than a neutral. I have merely protected realism against epistemology; and not denying epistemology.

And of course those epistemological prime things are *among* the things that there are—whether as classes of events of physical bodies, or whatever.

In relation to future chapters, function of present one has been negative: clearing epistemology out of the way, making ontology independent. But no realism is implied in the sense of an exclusive physical realism, a nominalism as regards abstractions. That is a later question. Prepared meanwhile to admit that maybe abstract objects—whether properties of phys. objects, or classes thereof, or whatever—may be needed; just as idealism might have needed classes or properties of sensa etc. besides the sensa themselves. Concrete vs. abstract is another cleavage, independent of subjective vs. objective, internal vs. external.

And still we aren't throwing out philosophy with epistemology, leaving ourselves with nothing but physics. There remains ontological problem of essentially philosophical character, though *not* epistemological—as we shall see.

A.3.2. Things and Events

When we proceed from a materialistic view to consider what there is, we are subject to the dictates of current physics, not to say geography, astronomy, etc. But we are concerned now with broad enough lines so that we needn't make a deep study of those things. Thus, we may take (subject to possible additions, but surely not subtractions) the spatially extended objects. These include common-sense physical objects. Also subtle conjectured physical objects, down to atoms and beyond. And the subtlest physics doesn't *deny* the tables & chairs, but only takes fuller a/c {account} of their parts. Our extended objects comprise the common-sense big ones *and* their subtle parts. (Let's not hear that the parts are different in kind, that's epistemology!) And this is a category which we can accept as a *sine qua non*, and which we can try riding to the exclusion of others until we see other needs and how they arise.

[Show now the paradoxes of time, e.g. "King George married a widow" and Heraclitus on the river.] Hence the advantages of a spatio-temporal approach; regions of space time: *events*, not things. Physics becomes *cosmography* (like geography, but including all the universe, and history). This is not Einstein's relativity, but a point of view which would have been useful long before.

5 Nov 1944[8]

Ontology is the philosophical discipline, or undiscipline, that decides what there is. But is it really the philosopher's business to decide this, or is it the scientist's? The fact is, as we shall see, that the question "What is there?" is broad enough to allow both philosopher and scientist to move about in it without treading on each other's toes.

31 Oct 1944[9]

What is there? There is everything. This answer, though incontestable, affords little satisfaction; it wants expanding. Thus we might take up the several things that there are and describe them, each in turn. In this way lies science as well as most of the other non-fictional departments of literature. The descriptive job is one that cannot be finished, but it has been carried a long way. The things that have been described individually are indeed rather limited: various land masses, seas, planets and stars, in the geography and astronomy books, and an occasional biped or other medium-sized object, in the biographies and art books. On the other hand, description has been put on a mass-production basis in zoology, botany, and mineralogy, where things are grouped according to similarities and described group by group. Physics carries mass description farther, admitting high generality by abstracting ruthlessly from difference in detail. Even the most general and speculative reaches of natural science are part and parcel of the descriptive answer to the question what there is, nor is pure mathematics to be

[8] Quine (November 5, 1944, item 3181). Although the present note was written a few days after the note that follows below, Quine intended it to precede that note. The present note has page number 1, whereas the notes that follow below encompass the pages 2–3 and 4–9 respectively.

[9] Quine (October 31, 1944, item 3181).

excluded; for the things about which the general question asks are in no way limited to concrete objects of the material world, to the exclusion of the numbers, sets, functions, etc. if such there be, whereof pure mathematics treats.

At least, they are not limited to concrete objects if there really *are* any other things, such as the alleged numbers, sets, functions, etc. whereof pure mathematics purports to treat. But *are* there? So we now find ourselves getting off into a philosophical phase of the question "What is there?", quite apart from the descriptive business that is the portion of the sciences and related subjects. We find ourselves in ontology.

5 Nov 1944[10]

The issue over abstract objects—not only numbers and kindred mathematical objects, but equally such alleged things as species and genes, diseases, laws, customs—is one among several ontological sore points. There are ontologies that admit abstract objects of one sort, known as properties, and repudiate others, known as classes; conversely, there are ontologies that admit classes and repudiate properties; there are ontologies that repudiate everything but the concrete, material things of the external world; and there are ontologies that challenge even these.

No matter where the truth may lie among these conflicting theories, the ontological question "What is there?" is a different question from the "What is there?" of unphilosophical science.

The natural scientist or mathematician begins with the unquestioning acceptance of all the alleged objects of one or another broad category—physical object, class, property—and proceeds to describe what there is in the category, individually or group by group; whereas the ontologist scrutinizes the claims to existence on the part of the members of the category—all physical objects, e.g.—as such. The difference turns on the broadness of the categories concerned. It turns on the difference between such special classifications as wombat, centaur, zombie, prime number, etc., on the one hand, and such general classifications as physical object, property, class, etc. on the other. Thus, given physical objects in general, the natural scientist is the man to decide about wombats and centaurs. Likewise, given classes, or whatever other realm of objects the mathematician needs, it falls to the mathematician, to decide whether in particular there are any even prime numbers or any cubic numbers which are sums of parts of cubic numbers. The scientist's initial uncritical acceptance of a realm of physical objects, or of {unreadable} classes, etc., is what is relegated to ontology for support or criticism.

The philosopher's task differs from that of the natural scientist or mathematician, no less conspicuously than the tasks of these latter two differ from

[10] Quine (November 5, 1944, item 3181).

each other. The natural scientist and the mathematician both operate within an antecedently accepted conceptual scheme but their methods differ in this way: the mathematician reaches conclusions by tracing out the implications exclusively of the conceptual scheme itself, whereas the natural scientist gleans supplementary data from observation of what happens around him. The philosopher, finally, unlike these others, focuses his scrutiny on the conceptual scheme itself. Here is the task of making things explicit that had been tacit, and precise that had been vague; of uncovering and resolving the paradoxes, smoothing out the kinks, lopping off the vestigial growths, *clearing* the ontological slums.

It is understandable, then, that the philosopher should seek a transcendental vantage point, outside the world that imprisons natural scientist and mathematician. He would make himself independent of the conceptual scheme which it is his task to study and revise. "Give me πoυ στω {a place to stand}," Archimedes said, "and I will move the world." However, there is no such cosmic exile. The philosopher cannot study and revise the fundamental conceptual scheme of science and common sense, without having meanwhile some conceptual scheme, whether the same or another no less in need of philosophical scrutiny, in which to work. The philosopher is in the position rather, as Neurath says, "of a mariner who must rebuild his ship on the open sea; he . . . ."

But if the philosopher has access to no transcendental vantage point, still his method differs in an important way from the methods of natural science and mathematics. Just what it is that distinguishes philosophical method is not generally agreed, and a complaint occasionally leveled[11] against the philosopher is that he has no method at all. I think, though, that the following, I venture, nonetheless, to suggest this as the characteristically, philosophical method: the making our fundamental terms more explicit and precise and deriving consequences from the increased explicitness and precision. This, it will be noted, is an enterprise quite compatible with Neurath's analogy of the ship at sea; it can be carried out without transcending the existing world view. Yet it is not to be confused with the task of empirical semantics, which inquires, as it were by a Gallup poll, into the meanings that people attach to the words they use.[*] What the philosopher asks is less what men have meant than what, for the purposes at hand, they

[11] Editorial correction of "levelled."

[*] Confusion of philosophical analysis with empirical semantics is occasionally to be encountered. Passages where G. E. Moore talks about "what people ordinarily mean by necessity" (check with Malcolm's paper) may come to mind, but there are clearer cases; see, e.g. the attack by the philologist. . . . {Alan} Gardiner upon the philosopher Bertrand Russell on the subject of proper names, in Gardiner, *The Thy. of Pr. Names,* also the informative and entertaining but philosophically irrelevant study by the polling psychologist Arne Næss on "The Concept of 'Truth' among those who are not professional philosophers," . . . {See Gardiner (1940) and Næss (1938). Quine's remark about Malcolm's paper probably refers to Malcolm (1940).}

might just as well have meant. If the philosopher is to introduce precision and order into what had been vagueness and chaos, the end-point of his interests is not to be sought in the chaos of vague meanings with which the fundamental terms are endowed by the uncritical man in the street. The philosopher wants a clear-cut set of fundamental *notions*, as free as possible of mystery and paradox, which will serve the purpose for which those vaguest, and most general terms are intended that form the substructure of unphilosophical science and uncritical common sense.

An appalling mass, certainly, of methodless. . . . However, there is a legitimate field and method of a type identifiable with philo.

A.4. An Extensionalist Definition of Meaning (1949)

Jan. 16, 1949

Dear Van:-

In the course of writing Morty last week I outlined a recent idea I had on a point all three of us have discussed, and asked him to turn it over to you for your comments.[12] The present letter is on a different, though related point. I have been assigning some of your articles as collateral reading in the course I am giving on my book, and in looking over "On what there is" I feel it is time to take issue on what I am sure is an argument[13] you have also used elsewhere. I am referring to the sentence on page 28 that reads: "But the two phrases cannot be regarded as having the same meaning; otherwise that Babylonian[14] could have dispensed with his observations and contented himself with reflecting the meanings[15] of his words."[16] Even if I grant you that that Babylonian spoke English and so used the phrases "Morning Star" and "Evening Star," I dispute your contention. The Babylonian could have dispensed with his observations all right if he knew all about the meaning of both terms; { . . . } he might use the words to some purpose, and the meanings of the words might be the same, without his knowing that those meanings are the same. Your argument depends on the premise: if x knows two words and if their meaning is the same, then x can find out without

[12] Unfortunately I have not been able to locate Goodman's letter to Morton White ("Morty") at either the Quine or the Goodman archives.

[13] Editorial correction of "anargument."

[14] Editorial correction of "Baylonian."

[15] Editorial correction of "meansings."

[16] See (OWTI, 1948, 9).

other information that their meaning is the same. This I[17] deny. You may argue that if x doesn't know the meanings of the two words he can't use them in framing questions and hypotheses; but I reply that x can know something about the meaning of two words without knowing all about them. He can discover later that they have the same meaning just as he discovers anything else; in fact there is no distinction between learning the meanings of words and learning anything else (apart from e.g. learning how to jump). When the Babylonian pulled out his trusty telescope (pre-Galilean model), he found out something and what he found out can be described either as something about the heavens or as something about the meanings of two words.

All this is quite obvious to an extensionalist, of course. Two terms have the same meaning if they have the same extension; if we know part of the extensions we know part of the meanings and can use the terms accurately, but even if the extensions are in fact the same we may never happen to know that. If you suppose that using the words at all correctly implies that one has their whole meanings stuck in his navel for inspection, you get ideas about meanings being ideas—and that gets you into all sorts of[18] trouble. Why not be an extensionalist again?

You might in turn show this to Morty, if you will.

And I'll[19] {remainder of the letter missing}

January 27, 1949
Professor Nelson Goodman
Schwenckville, Pennsylvania
Dear Nelson,

I object when you say that it is quite obvious to an extensionalist that two terms have the same meaning when they have the same extension. It is as if you were to say that it is quite obvious to an extensionalist that two terms have the same rhythm (or length, or derivation) when they have the same extension. On the contrary, it is no breach of extensionalism to accord terms all manner of features (rhythm, length, derivation, meaning) over and above their extensions.

The extensionalist will merely insist that the rhythms, lengths, derivations, and meanings themselves either (a) not really exist, the sensory references to them being explained away contextually, or else (b) be individuals, or else (c) be classes of individuals, or (d) classes of such classes, or (e) etc. Alternatives (c), (d), and (e), of course, are open only to those extensionalists who are also platonists.

[17] Editorial correction of "i."

[18] Editorial correction of "sortsof."

[19] Editorial correction of "And I8ll."

There is another doctrine, not to be confused with extensionalism (whereof we are both adherents), which might be named *asemiotism.* It is the doctrine that no coherent notions of meaning, analyticity, etc. are to be hoped for. While I do not identify myself with asemiotism, I must concede it a certain plausibility. I used to think of you, indeed, as one of its staunchest proponents; and I urge you to reconsider your recent abandonment of the doctrine. For you do evidently abandon it; far from repudiating the concept of meaning as a good asemiotist should, you now espouse a theory of meaning according to which the meaning of a name is the thing it names and the meaning of a general term is its extension. This, far from being asemiotism, is merely a brand of neo-Confusionism.

Once you tackled the problem of contrafactual conditionals, pointed out the difficulties, and, in the end, considered that we have not made, maybe can not make, sense of contrafactuals. You didn't draw the conclusion that therefore contrafactual conditionals are simple material conditions. This I applaud. My policy in the theory of meaning is the same. From the fact that "meaning" has not been or cannot be made to make sense, let us not conclude that it therefore does make sense and that that sense is the named object or the extension. Or (which is the parallel inference) that analyticity is truth.

In your letter to Mortie you hinted an opposite, equally untenable resolution: the meaning of a term is the term, or word, itself. You showed ingeniously that difference of meaning (or picture, anyway) and difference of word are only a matter of degree. I, however, a believer in matters of degree, am not led from the fact that a difference is a matter of degree to the conclusion that the difference is an identity.

Certain "conditions of material adequacy" (in Tarski's phrase) govern any non-Pickwickian notion of meaning. One is that if x understands two words (the phrase "x knows two words," in your letter, harbors a tendentious amphibole), then he knows their meanings and can determine whether they have the same meaning by reflecting on their meanings. This I use in the passage on p. 28 of "On what there is," to which you object. I maintain the following law of trichotomy: we must either subscribe to the above, or eschew the word "meaning," or be wantonly misleading in our use of the word.

For me the phrase "knows-the-meaning-of" is quite interchangeable with "understands," and the one is as clear as the other. If meanings are to be singled out as entities, or if the word "meaning" is to be extended to other contexts than "knows-the-meaning," it remains a fundamental condition of the problem, to my mind, that "knows-the-meaning-of" remain "alike-in-meaning" with "understands."

Pending a satisfactory definition of meaning (supposing such a theory even possible), I cannot of course, use the notion as a basis for my own philosophical constructions, insofar anyway as those are to be clear and meaningful. But what happens in places like the aforementioned page 28 (and again in my review of Everett Nelson)[20] is quite different: I am using the opponents' "meaning" jargon conditionally against him in polemic, reducing him to absurdity in his own terms. One's case is far stronger in this form than when rested on the doubtful postulate of eternal asemiotism.

Perhaps your advice in[21] such cases is "leave such people alone." But this is undesirable because of the effects false doctrines of meaning have *outside* themselves—e.g., upon ontology, through confusions of meaning with naming and hence of meaningfulness with namingfulness. This, in fact, is how the matter enters my paper "On what there is."

Yours,

P.S. The department subsidy for your book can be used elsewhere than Harvard Press if you like.

A.5. The Present State of Empiricism (1951)

I am much flattered that my recent paper "Two dogmas of empiricism" should have been chosen as basis for today's discussion. But I do not flatter myself that the paper contributes a new idea to philosophy. The paper is negative: an expression of distrust of two doctrines which have come to assume a central position in current empiricism. I do not reject empiricism, but in rejecting the two component doctrines in question I argue for an increased sympathy with the pragmatists and, in certain respects perhaps, with the idealists.

The two dogmas concerned are the doctrine of analytic statements and the doctrine of reductionism. The latter may be regarded as the more basic of the two, and I shall review my position primarily with regard to it. So the issue becomes an issue between those who, with me, advocate rejection of reductionism, and those who would rather see some reduction of rejectionism.

The reductionism which I reject is, in its radical form, the doctrine that all significant statements are reducible one by one to protocol language, language about experience. Actually this is not a doctrine, but a two-dimensional space of doctrines. One of the two independent variables is our standard of what is to qualify as protocol language, and the other independent variable is our standard of what is to pass for reduction. Reductionism in its extreme form takes the protocol language as comprising just the devices of elementary logic, plus predicates which express qualities and relations given in immediate private

[20] Quine (1947).

[21] Editorial correction of "is."

experience; and reductionism in its extreme form takes *reduction* as consisting in out-and-out translation. In more moderate forms of reductionism, the protocol language is imagined as liberalized; and also the type of reduction insisted upon is slackened to include inferential connections of a weaker sort than translational equivalence. But there remains, throughout even these more moderate forms of reductionism, a common trait: there remains the doctrine that each significant statement has its own separate and peculiar empirical content, its own separate and peculiar connections with immediate experience. Now I regard even this attenuated form of reductionism as groundless and implausible.

Reductionism at its most sanguine is represented by Carnap's *Logischer Aufbau der Welt.*[22] Carnap later despaired of the program envisaged in the *Aufbau.* He shifted his ground from phenomenalism to physicalism. Moreover he relaxed his standards of *reduction* so as to allow for reduction forms of the kind described in *Testability and meaning,*[23] which stop short of actual definition or translation. In Carnap's current phase, his confirmation-theory phase, we find reductionism at its least assertive. At this stage, reductionism has been so watered down that all that needs to be accounted for any longer is a relation of confirmation between a scientific statement and the relevant observations.

But reductionism even thus diluted is not dilute enough to suit my tender palate. The confirmation relation remains, significantly, a pretty mysterious relation. Intuitively, confirmation of a statement by an experience is felt to consist in an increase of the probability of the statement; but the notion of the probability of a statement is a notoriously obscure and baffling notion, once we take our leave of cozy pseudo-universes consisting of black and white balls in an urn.

What I urge is still further dilution of reductionism; dilution beyond recognition. This further step which I urge consists in denying that the statements of science, taken one by one, have any empirical content at all. I suggest that it makes no sense, except as a rough abstraction occasionally feasible in simple cases, to speak of even the partial confirmation of a statement by an experience. I suggest that what has empirical support, what is capable of fitting experience or conflicting with experience, is not the isolated scientific statement, but only the systematic network of scientific statements as a totality.

A science which engenders a false expectation of experience, a false prediction, is faulty and needs revision; but the fault cannot in principle be localized in one component statement of the system rather than another, and the fault can be mended by making any of various alternative adjustments at various points of the system.

[22] Carnap (1928a).

[23] Carnap (1936; 1937).

What I am arguing is thus rather reminiscent of the coherence theory of the idealists. But I do not reject empiricism, for I do still maintain that the scientific system as a totality has more than the requirements of internal simplicity to answer to; the system as a whole continues to face the tribunal of sense experience. The system as a whole may be viewed, indeed, as an apparatus for anticipating sensation.

This holistic empiricism is not new. The same attitude was emphasized e.g. by Duhem, who urged that "an experiment in physics can never condemn an isolated hypothesis, but only a whole theoretical conjunct . . . The crucial experiment is impossible in physics." In Einstein's phrase, "we are seeking for the simplest possible system of thought which will bind together the observed facts."[24]

These are commonplace remarks, so commonplace that one tends to overlook the opposition in which they stand to other commonplace remarks such as Peirce's maxim that the meaning of *a statement* consists in the difference which the truth or falsity of *that statement* would make in future experience.

It may seem that Peirce's maxim can be reconciled with the holistic view in this way: we can construe Peirce as meaning that the meaning of a statement consists in the difference which the truth or falsity of that statement would make in future experience *supposing* all other statements to be held fast in point of truth value. But a little reflection shows that this exegesis fails. If you vary the truth value of one statement, presumably you should concomitantly vary the truth values of other statements which are logically implied by the given statement, or which logically imply it. Also presumably you will want to vary other statements which are connected with the given one by so-called well-established empirical laws. In short, I see no way of making acceptable sense of Peirce's

[24] In the margin, Quine writes: "La theorie physique ('16). The World as I See It." See Duhem (1914) and Einstein (1935). Quine did not refer to Duhem in "Two Dogmas of Empiricism." Five days before he presented "The Present State of Empiricism," however, Peter Hempel sent a letter to Quine in which he explained that Duhem also defended a variant of holistic empiricism:

> A propos of your statement that the statements of empirical science face the tribunal of direct experiential evidence not individually, but jointly, I recalled Frank as saying (and ditto Dubislav in his Naturphilosophie as mentioning) that Pierre Duhem had taken a very similar position. I looked again into his La théorie physique (1906) and found this confirmed, and I thought you might be interested in the sixth chapter of the second part. (May 21, 1951, item 499)

The reprint of "Two Dogmas" in *From a Logical Point of View* (1953) also contains a footnote citation of Duhem.

maxim, which addresses itself to isolated strands of the scientific fabric rather than to the fabric as a whole.

Peirce has been called a pragmatist, but his maxim is, we see, simply the mild form of the statement-by-statement reductionism which I reject. James' pragmatism lies closer to the holistic view which I urge. For I urge that science as a whole is an apparatus for anticipating sensation, and that it is to be evaluated and modified from day to day with an eye simply to its pragmatic value relative to the anticipation of sensation. Internal simplicity and intuitiveness are of course desirable in scientific systems, but the pragmatic value of these features also is evident.

I do not go along with the pragmatic sanction of the will to believe. I must emphasize that my pragmatism limits itself to gauging science by its instrumentality toward one express end only; that end is not happiness, not comfort, but anticipation of sensation.

As long as it seemed reasonable to speak of the empirical content of an individual statement, and the rules of confirmation of an individual statement, it seemed reasonable also to countenance *extreme* cases which were confirmed *ipso facto* come what may. Here we have the notion of an analytic statement. But in "Two dogmas of empiricism" I have argued that the various known ventures at explicating the notion of analyticity fail of their purpose, and I have suggested that the very effort to draw a distinction between the so-called analytic and the so-called synthetic is a false lead in theory of knowledge. Certainly the supposed distinction seems much less urgent and much less plausible when we withdraw to a holistic empiricism.

For, what is the function of a typical statement of allegedly analytic kind, say a truth of logic or mathematics? Logic and mathematics are, we may all agree, useful as auxiliaries to so-called empirical science. They are portions of the total scientific scheme which are relevant to experience only indirectly through the workings of the scientific scheme as a whole. But, from the point of view of holistic empiricism, the same can be said not only of the truths of logic and mathematics but of science generally; so that the conspicuous motive for setting the so-called analytic statements over against the so-called synthetic ones disappears.

There then remains no basic difference in kind between the laws of logic or mathematics and the hypotheses of physics; or between the hypotheses of physics and the decisions of ontology, such e.g. as the decision to admit classes or physical objects within the range of our variables of quantification.

It is particularly to be noted that the question whether there are hippopotami, the question whether there are electrons, the question whether there are external objects, and the question whether there are classes, come to be in principle alike. In each case the alleged objects are not to be sought in constructs from sense data. In each of these cases, rather, the decision whether to *posit* the alleged objects rests on questions of the overall simplicity of the resulting science plus the question of its squaring at its edges, so to speak, with sense experience. In "Two dogmas" I drew an analogy between the positing of physical objects to simplify the systematization of sense experience, and the positing of irrational numbers to simplify the systematization of rational numbers; this analogy, I now find, was drawn much earlier by Prof. Frank.[25]

I do not ignore the evident difference in seeming proximity of sensory verification between the statement that there are hippopotami and the statement that there are electrons or classes. But I think the difference arises as follows. In general, when a recalcitrant experience requires us to change our system, we are guided, I think, by *two forces* in choosing where and how to revise the system. One force is the quest of simplicity; the other is conservatism. Now certain statements, such as that there are or are not hippotami or that there are brick houses in Elm Street, are likely to be seized upon for revision in preference to other portions of science, when certain recalcitrant experiences arise; and the reason for such preference is that we see better how to preserve the totality of the system relatively undisturbed by revising these than by revising others. Statements such as that there are classes, on the other hand, are so to speak more central to the conceptual scheme—by which I mean merely that we see no similar reason to revise *these* statements rather than others, in readjusting our science to any special recalcitrant experiences.

We are led thus to a vague notion of relative centrality, which somewhat takes the place of the absolute of analyticity. Now this notion of relative centrality is indeed unsatisfactorily vague, and I should hope for some sharpening of this and kindred notions as the theory of knowledge progresses. Possibly the device of an artificial language or the idea of an ideal language, which has figured in varying ways in the work of Carnap, Bergmann, and Sellars, will play a rôle in such future progress.

On the other hand I have misgivings about preserving the old notion of analyticity, or the old notion of the empirical content of a statement, and I have misgivings about erecting ideal systems around *these particular* notions; for I suspect that these notions may actually have been impeding our progress,

[25] On April 1, 1951, Frank had written Quine that "[i]ncidentially the example of the rational and irrational numbers which is to illustrate the distinction between objects and sense impressions was used by me in a paper of 1929 to a very similar purpose." Quine responded four days later that he greatly regrets "not having known, so as to have made proper citations" and that he "shall make an opportunity, in my next relevant publication to make amends" (item 370).

standing in the way of what may prove to be more illuminating epistemological conceptualizations.

Response to Bergmann

1) {I am} not committed to viewing math. {&} logic as psychology or sociology of discovery. No more so than physics; for I have been minimizing that boundary. *Epistemology* comes closer than math-logic to psych. or sociology of discovery.
2) I haven't thought of my view as entailing a broader or more Aristotelian ontology than other views entail. In my view there is a premium on economy & simplicity of conceptual scheme, & ∴ {therefore} on narrowness of ontology.
3) Carnap's extension of truth-table sort of analyticity to *general* logic (idea of identical formula) can be continued into any field we like, by Tarski's method of recursive def'n of truth.
4) But I grant that this is unnecessary for B's purpose, *if* he can construe *descriptive* versus *logical* constant. For a logical truth can (as emphasized in "2 dogmas") be defined in terms of descript. const. & truth.
   If we assume ideal lang. we can define logical const. by enumeration & descriptive const. by subtraction.
   Process presupposes what proceeds? No. Time as a 4th dimension.
5) Passim—Don't let me discourage construction of the ideal language (supplying the prim. preds. & defs.). Until it is done, please excuse my unsanguineness. When it is done, I'll be as pleased as you. But there is many a slip bet. cup & the lip; or better, there is many a catch between blueprint & sketch.
6) I don't say blueprint or nothing, except in cases where I feel real doubt as to whether the sketch is capable in principle of being filled in to approach blueprint even asymptotically. (Nagel too.)

Response to Von Mises

DOCTRINE 1: Whether a stmt. {statement} is "taut." {tautological} is indep. of the lang. in which it occurs. Nobody believes this.

DOCTRINE 2: Given the lang., say a limited technical language, statements can be classified 1 by 1 *within that lang.* This Carnap believes; this von M. believes; and *this* is what I throw into question. Prof. v. M. appeals to the notion of *axiom* to buttress this notion; my criticism of this approach would be the same as my crit. of the device of *semantical rule.*

Then he speaks of whole *systems* as taut., but this seems to be merely another word for deductive systematizations (of any desired doctrine).

On other points he seems to share my view, exc. for some diff'ce of usage of the term reductionism.

Response to Feigl

Really abbreviative def'ns can't matter so much for the need of a distinction bet. analytic & synthetic; for, insofar as the def'ns are really abbreviative, they are superfl.

I grant there is a rule-like component of language & a factual component, in this sense: truth depends both on lang. & sens'n.; better, systems are in language, & have to square with sensations. But can't carry the split of components into the individual stmts., even considered in relation to the whole.

I might be expected to draw comfort from F's. functional analyticity, i.e. his concession that the distinction of analytic & synthetic is systematically ambiguous, & varies with oscillations in meaning or attitude. But I don't. For I still don't know what this varying distinction is; I don't know what is varying & ∴ {therefore} can't appreciate the variation.

He says in each alternative the notion of analyticity keeps its pristine clarity. I wonder how much more this means than that each dichotomy of the class of truths will, if identified with the analytic-synthetic distinction, constitute one possible & clear analytic-synthetic {remainder of the sentence unreadable}.

Response to Bergmann

Beyond my strictures of this {morning}, I have little to say beyond agreement. B has stated our diff'ces much to my satisfaction: that I can agree to an enumerative dist'n bet. logical and descriptive signs, but can't agree to giving it epis. signif. for this would amount to granting the analytic-synth.

Right also in relating this to my refusal to distinguish in principle bet. "There is a drugstore around the corner" & "There are universals." Carnap e.g. would put ontological questions on the side of choice of language system, & drugstore q'ns on the side of q'ns. of fact. I would rub out the line. Instructive to see that the prob. of analyticity & the prob. of the nature of ontological issues are one.

Response to Feigl

I want to stress the diff'ce bet. a fabric & a monolith. I've likened science to a fabric, also to a field of force; but wouldn't liken it to a monolith. Scientific progress proceeds precisely by modifying parts. Nagel says you don't start by setting up a system. Granted; you are born into it.

System as a whole only in that we *can* do it here *or* here. Cohen & Nagel! Frank's answer good. Bar Hillel is answered. I've never spoken of abandoning a system & putting in a new system.

What F. wants here is, he himself suggests, provided perhaps by considerations of centrality. But I deny an absolutely fixed part, fixed differently in principle from other parts that happen to be preserved. I believe with Neurath. . . .

As for many-valued logic as a help for quantum theory, I have taken no stand; it is enough for my example that such a proposal has been made. And such have been made ≥4 times.

On uniqueness of el. logic, a predicament: new mngs. for old words vs. new doctrines about old mngs. *Cooley*! Could say same in any {unreadable}. Can't sep. mng. from fact!

Response to Margenau

How literally do I hold that the business of science is anticipation of exp'ce?

Don't strongly object to Bridgman's & M's wish to include past conformities, except insofar as time & memory are themselves archaic posits or components of the conceptual scheme. Certainly there are problems here. What I am trying to do is give to the conceptual scheme everything except the raw confirmatory experiences, & to find the external *purpose of* the conceptual scheme in those experiences.

Response to Nagel

N. asks whether I am claiming difficulty in analytic-synthetic only for unformalized languages; ans. is *no*, & is continued in "2 dogmas," where I emphasize that the problem has the same difficulty in artificial langs. This was where I talked about Carnap's semantical rules & his theory of ranges.

N. is right that we might say arbitrarily what we are going to call *analytic-in-*L_0, . . . L_1, etc.; this I remarked in "2 Dogmas"; but doesn't help for *analytic*, i.e. *analytic-in-L* (vbl.), even for an artificial L.

If we can get *sentence mngs.* (which I doubt), I expect one can abstract *fragment mngs.* (e.g. for logical connectives) satisfactorily for N's. purposes; not by appeal to axioms as fund'l. (this would be circular, presupposing "analytic" or "axiom" or "semantical rule"), but by appeal to substitution classes of expressions (as linguists now tend).

N. shares my rejection of reductionism; so do most of the other members of the panel. Seems to me, as I argued this morning, that this is enough to remove the main motive & plausibility of analytic.

Response to Wang

I am all for little red'ns; every new posit which is sidestepped by a short red'n is a gain; Occam's razor. I like red'n of math. to set thy., & I'*d* like red'n of disposition terms to their cspdg. {corresponding} root terms.

On deriv'n of analyticity from centrality, no objection—once we have made fair sense of centrality, & *if* we find this explic'n of analyticity compares well enough with past philosophical usage to make the name appropriate.

Response to Bar-Hillel

I have no quarrel with dichotomies. Prime vs. composite O.K; Animal vs. veg. O.K. (despite vagueness); front & back of head O.K.; but not clockwise.

I thought I reduced the special plea for language systems to absurdity in "2 dogmas" by saying: yes, define a lang. system as an ordered pair of a lang. & a class of stmts. so-called analytic for that lang.; & then *define* (in *general* for *all* language systems!) analytic for a lang. system as membership in 2d component! Or better still, let's stop tugging at our bootstraps altogether.

I grant we'd have an explication of synonymy & analyticity if we could get his triadic equisignificant, in behavioristic terms. This is what I'd regard as a solution, precisely. [(Morris too.) I've said so often in lectures & probably in print, briefly. I'm all for psychologizing. I've worked on it.] This is what I'm not hopeful of.* I think B. H. has helped bring out the detailed character of the problem.

Response to Frank

Say regarding his remarks, as regarding von Mises', that I don't repudiate logical consequence. But I do doubt the basic distinction which Fr. draws between the certainty of logical consequence & the uncertainty of so-called empirical statements. Here I expect differences of degree, something like degree of centrality (if this notion should evolve into a less vague status); but the complete & fund'l diff'ce which F. imagines here, if he does imagine one, is what I question in questioning the boundary bet. analytic & synthetic.

On F's remarks on red'ism I am in agreement, stressing that the deductions to sense expce. are from the whole system & not sentence by sentence. Even the rules of deduction, the logical rules, are part of the system which as a whole is brought against expce.

* Something like centrality instead, but à la Morris & Bar-Hillel.

F. certainly right & interesting, in what he says about my rejecting a later phase of log. empiricism. I recognize I am in closer sympathy with Neurath & Duhem on these points. I am doubtful about Poincaré; but this historical question may be passed over. In any case, the more confirmation I can find in earlier thinkers, the more secure I feel.

A.6. The Sophisticated Irrational (1965)

Trainboard, Dorset, July 21, 1965

Certain hopeful positivistic tenets turned hopeless (*Aufbau* reduction; tabula rasa; verif'n thy. of mng.; protocol sentence, in some forms; atomic facts & cspdce. {correspondence}; analyticity as a way of immunizing math.).

Valuable supervening insights: We must work within a growing system to which we are born (involutionism) (Popper, Neurath). [Conservatism follows (Popper; oddly Neurath not).] We can't gen'ly apportion empirical content to statements separately. It is a synchronic & diachronic contextualism. ("Holism" is too strong.)

But they are carried away by the mood of iconoclasm. They forget what they came for. Like the invading army destroying conquered riches. Or our own commissars of urban-renewal, destroying houses in the name of housing shortage.

Thus Hanson & Gestalt psych. against observation. Failing to see that involutionism & behaviorism take care of obsvn. sentences to perfection. Don't need to care about subjective awareness or time-consuming acts of inference or their absence. Sentences varying about the observations as Hanson describes them simply aren't obsvn. sentences for a community embracing the variations. (Popper is wrong here too.)

On Kuhn. Good on involutionism. Adds illuminating stuff on culture-jumps, too. But is carried away by cultural relativism. No truth. Involutionism (immanentialism) carries the cure for this. Tarski. Popper O.K. here.

Mustn't Kuhn feel scientists have got somewhere? Doesn't he feel puzzled by that?[26] Solution above.

Oxford nihilism regarding ontology & epistemology.

Are there no problems? Nothing to do? Outline. 1) Rel'n of simplicity to pred'n.

[26] In his copy of Kuhn's *The Structure of Scientific Revolutions* (1962), Quine writes in the margin of page 172: "In questioning theoretical progress and passing over the concept of truth he seems to wholly ignore the fantastic phenomenon of man's technological progress. Radio; aviation; nuclear power; orbiting. What does he want—proof?" (Quine's library, item AC95.Qu441.Zz962k).

BIBLIOGRAPHY

Archival Sources

W. V. QUINE PAPERS

The unpublished papers, letters, lectures, and notebooks listed are stored at the Harvard Depository and can be accessed at Houghton Library. A catalog of Quine's unpublished work can be found at http://oasis.lib.harvard.edu/oasis/deliver/~hou01800. The references are ordered by item number.

Quine, W. V. & American Philosophical Association (1936–1986). Correspondence. MS Am 2587: Box 001, Item 31.
Quine, W. V. (various dates). Correspondence Ap- through As-. MS Am 2587: Box 002, Item 40.
Quine, W. V. & Bergström, L. (1988–1996). Correspondence. MS Am 2587: Box 003, Item 86.
Quine, W. V. & Beth, E. W. (1947–1964). Correspondence. MS Am 2587: Box 004, Item 96.
Quine, W. V. & Center for Advanced Study in the Behavioral Sciences (1955–1979). Correspondence. MS Am 2587: Box 007, Item 205.
Quine, W. V. & Church, A. (1935–1994). Correspondence. MS Am 2587: Box 008, Item 224.
Quine, W. V. & Clark, J. T. (1951–1953). Correspondence. MS Am 2587: Box 008, Item 231.
Quine, W. V. (various dates). Correspondence Co-. MS Am 2587: Box 009, Item 234.
Quine, W. V. & Columbia University (1949–1970). Correspondence. MS Am 2587: Box 009, Item 248.
Quine, W. V. & Conant, J. B. (1951–1979). Correspondence. MS Am 2587: Box 009, Item 254.
Quine, W. V. & Cooley, J. C. (1932–1962). Correspondence. MS Am 2587: Box 009, Item 260.
Quine, W. V. & Creath, R. (1977–1998). Correspondence. MS Am 2587: Box 009, Item 270.
Quine, W. V. & Davidson, D. (1957–1997). Correspondence. MS Am 2587: Box 010, Item 287.
Quine, W. V. & De Laguna, G. A. (1950–1954). Correspondence, MS Am 2587: Box 010, Item 293.
Quine, W. V. (various dates). Correspondence Di- through Do-. MS Am 2587: Box 011, Item 306.
Quine, W. V. & Dreben, B. (1948–1997). Correspondence. MS Am 2587: Box 011, Item 315.
Quine, W. V. (various dates). Correspondence Er- through Ez-. MS Am 2587: Box 011, Item 336.
Quine, W. V. & Frank, P. (1951–1967). Correspondence. MS Am 2587: Box 013, Item 370.
Quine, W. V. & Goodman, N. (1935–1994). Correspondence. MS Am 2587: Box 016, Item 420.
Quine, W. V. & Harvard University, Department of Philosophy (1930–1994 and undated). Correspondence. MS Am 2587: Box 018, Item 471.
Quine, W. V. & Harvard University, Faculty of Arts and Sciences (1931–1988). Correspondence. MS Am 2587: Box 018, Item 473.
Quine, W. V. & Harvard University, Grants (1941–1988). Correspondence. MS Am 2587: Box 018, Item 475.

Quine, W. V. & Harvard University, President's Office (1937–1998). Correspondence. MS Am 2587: Box 018, Item 479.
Quine, W. V. & Hempel, C. G. (1936–1997). Correspondence. MS Am 2587: Box 019, Item 499.
Quine, W. V. & Hookway, C. (1988). Correspondence. MS Am 2587: Box 020, Item 529.
Quine, W. V. & Horwich, P. (1991–1992). Correspondence. MS Am 2587: Box 020, Item 530.
Quine, W. V. & Institute for Advanced Study (1955–1965). Correspondence. MS Am 2587: Box 020, Item 545.
Quine, W. V. (various dates). Correspondence Ir- through Iz-. MS Am 2587: Box 021, Item 553.
Quine, W. V. & Journal of Symbolic Logic (1936–1996). Correspondence. MS Am 2587: Box 021, Item 570.
Quine, W. V. & Lakatos, I. (1964–1974). Correspondence. MS Am 2587: Box 023, Item 616.
Quine, W. V. & Leonelli, M. (1966–1998). Correspondence. MS Am 2587: Box 024, Item 637.
Quine, W. V. & Lewis, C. I. (1929–1996). Correspondence. MS Am 2587: Box 024, Item 643.
Quine, W. V. & Margolis, J. (1967 and undated). Correspondence. MS Am 2587: Box 025, Item 675.
Quine, W. V. & Miller, H. (1948–1952). Correspondence. MS Am 2587: Box 027, Item 724.
Quine, W. V. & Morris, C. W. (1936–1947). Correspondence. MS Am 2587: Box 027, Item 741.
Quine, W. V. & Myhill, J. (1943–1985). Correspondence. MS Am 2587: Box 028, Item 755.
Quine, W. V. & Nagel, E. (1938–1964). Correspondence. MS Am 2587: Box 028, Item 758.
Quine, W. V. & Putnam, H. (1949–1993). Correspondence, MS Am 2587; Box 31, Item 885.
Quine, W. V. & Rockefeller Foundation (1945–1980). Correspondence. MS Am 2587: Box 033, Item 921.
Quine, W. V. & Rodriguez-Consuegra, F. (1987–1994). Correspondence. Ms Am 2587: Box 033, Item 923.
Quine, W. V. (various dates). Correspondence Sc-. MS Am 2587: Box 034, Item 958.
Quine, W. V. & Sellars, W. (1938–1980). Correspondence. MS Am 2587: Box 034, Item 972.
Quine, W. V. & Skinner, B. F. (1934–1998). Correspondence. MS Am 2587: Box 036, Item 1001.
Quine, W. V. & Smart, J. J. C. (1949–1998). Correspondence. MS Am 2587: Box 036, Item 1005.
Quine, W. V. & Sosa, E. (1970–1995). Correspondence. MS Am 2587: Box 036, Item 1014.
Quine, W. V. & Weiss, P. (1937–1972). Correspondence. MS Am 2587: Box 042, Item 1200.
Quine, W. V. & White, M. (1939–1998). Correspondence. MS Am 2587: Box 043, Item 1213.
Quine, W. V. & Williams, D. C. (1940–1994). Correspondence. MS Am 2587: Box 043, Item 1221.
Quine, W. V. & Woodger, J. H. (1938–1982). Correspondence. MS Am 2587: Box 044, Item 1237.
Quine, W. V. & World Congress of Philosophy (1952–1998). Correspondence. MS Am 2587: Box 044, Item 1239.
Quine, W. V. & Wurtele, M. G. (1938–1997). Correspondence. MS Am 2587: Box 044, Item 1244.
Quine, W. V. (1950–1959). Correspondence concerning requests to publish or for copies. MS Am 2587: Box 045, Item 1263.
Quine, W. V. (1960–1982 and undated). D. Reidel Publishing Company, Editorial correspondence. MS Am 2587: Box 049, Item 1355.
Quine, W. V. (1939–1940). *Mathematical Logic*. Editorial correspondence. MS Am 2587: Box 050, Item 1391.
Quine, W. V. (1947–1950). *Methods of logic*. Editorial correspondence. MS Am 2587: Box 051, Item 1401.
Quine, W. V. (1968–1988). *Ontological relativity*. Editorial correspondence. MS Am 2587: Box 052, Item 1422.
Quine, W. V. (1956–1959). *On translation*. Editorial correspondence. MS Am 2587: Box 052, Item 1423.
Quine, W. V. (1948–1975). On what there is. Editorial correspondence. MS Am 2587: Box 052, Item 1443.
Quine, W. V. (1946–1949). *Theory of deduction*. Editorial correspondence. MS Am 2587: Box 053, Item 1467.
Quine, W. V. (1952–1960). *Word and object*. Editorial correspondence. MS Am 2587: Box 054, Item 1488.
Quine, W. V. (1959–1980 and undated). *Word and object*. Editorial correspondence. MS Am 2587: Box 054, Item 1489.

Quine, W. V. (1966–1974). *Words and objections.* Editorial correspondence. MS Am 2587: Box 054, Item 1490.

Quine, W. V. (undated). Quine's annotated copy of Putnam's *Meaning and the moral sciences.* MS Am 2587: Box 119, Item 2388a.

Quine, W. V. (1968). Epistemology naturalized; or, the case for psychologism. Typescript with manuscript additions. Prepared March 1968. Presented at The New School of Social Research, April 28, 1968. MS Am 2587: Box 085, Item 2441.

Quine, W. V. (1993). In conversation: Professor W. V. Quine. Interviews by Rudolf Fara. Printout with autograph manuscript revisions. MS Am 2587: Box 086, Item 2498.

Quine, W. V. (1966–1967). Russell's ontological development. Autograph manuscript, typescript carbon. Presented at a symposium of the American Philosophical Association, June 29, 1966. MS Am 2587: Box 090, Item 2733.

Quine, W. V. (1965). Stimulus and meaning. Autograph manuscript. Prepared September 11–25, 1965. MS Am 2587: Box 091, Item 2756.

Quine, W. V. (1946). *A short course in logic.* Typescript mimeograph. MS Am 2587: Box 097, Item 2829.

Quine, W. V. (1948). *Theory of deduction.* Typescript mimeograph. MS Am 2587: Box 097, Item 2830.

Quine, W. V. (1979). Foreword to the third edition of *From a logical point of view.* Autograph manuscript. MS Am 2587: Box 098, Item 2836.

Quine, W. V. (1986). Questions for Quine by Stephen Neale. Autograph manuscript, Manuscript, Typescript. MS Am 2587: Box 101, Item 2884.

Quine, W. V. (1968). For Rockefeller lecture. Autograph manuscript. Presented at Rockefeller University, February 5, 1968. MS Am 2587: Box 101, Item 2902.

Quine, W. V. (1968). For symposium with Sellars. Autograph manuscript. Presented at the Summer Institute of Philosophy, Southampton, Long Island, July 11, 1968. MS Am 2587: Box 101, Item 2903.

Quine, W. V. (1987). The inception of "New foundations." Autograph manuscript. Presented at Oberwolfach, March 1987. MS Am 2587: Box 102, Item 2928.

Quine, W. V. (1967). Kinds. Autograph manuscript. Prepared September 22, 1967. Presented at Brooklyn College, October 17, 1967. MS Am 2587: Box 102, Item 2948.

Quine, W. V. (1982). Levine seminar questions for Quine. Autograph manuscript, Typescript. MS Am 2587: Box 102, Item 2952.

Quine, W. V. (1940). Logic, math, science. Typescript. Presented at the Harvard Logic Group, December 20, 1940. MS Am 2587: Box 102, Item 2954.

Quine, W. V. (1950). Mathematical entities. Typescript. Presented at New York University, November 26, 1950. MS Am 2587: Box 102, Item 2958.

Quine, W. V. (1937). Nominalism. Autograph manuscript. Presented at the Harvard Philosophy Club, October 25, 1937. MS Am 2587: Box 103, Item 2969.

Quine, W. V. (1970). The Notre Dame lectures. Autograph manuscript. Presented at the University of Notre Dame, May 11–15, 1970. MS Am 2587: Box 103, 2971.

Quine, W. V. (1967). Ontological relativity. Typescript. Prepared March 1967, presented at the University of Chicago and Yale University, May 1967. MS Am 2587: Box 103, Item 2994.

Quine, W. V. (1968). Ontological relativity. Typescript with autograph manuscript revisions. MS Am 2587: Box 103, Item 2995.

Quine, W. V. (1952). The place of a theory of evidence. Typescript. Presented at Yale University, October 7, 1952. MS Am 2587: Box 104, Item 3011.

Quine, W. V. (1951). The present state of empiricism. Autograph manuscript. Presented at the American Academy of Arts and Sciences, May 26, 1951. MS Am 2587: Box 104, Item 3015.

Quine, W. V. (1967). Word and object seminar. Autograph manuscript. Presented at Ohio State University, January 9–13, 1967. MS Am 2587: Box 106, Item 3102.

Quine, W. V. (1953). Philosophy 148. Typescript. MS Am 2587: Box 107, Item 3158.

Quine, W. V. (1937–1944). Early jottings on philosophy of language. Collection of notes. MS Am 2587: Box 108, Item 3169.

Quine, W. V. (various dates). Erledigte notizen. Collection of notes. MS Am 2587: Box 108, Item 3170.

Quine, W. V. (1944–1951). Ontology, metaphysics, etc... Collection of notes. Box 108, Item 3181.
Quine, W. V. (various dates). Philosophical notes. Collection of notes. Boxes 108 and 109, Item 3182.
Quine, W. V. (1951–1953). Pragmatism, etc.... Collection of notes. Box 109, Item 3184.
Quine, W. V. (1925–1931). Miscellaneous papers. Collection of authograph student papers. MS Am 2587: Box 110, Item 3225.
Quine, W. V. (1930–1931). Papers in philosophy. Collection of authograph student papers. MS Am 2587: Box 111, Item 3236.
Quine, W. V. (1934). General report of my work as a Sheldon Traveling Fellow 1932–1933. Typescript. MS Am 2587: Box 112, Item 3254.
Quine, W. V. (ca. 1947). Philosophy 148. Autograph manuscript. MS Am 2587: Box 115, Item 3266.
Quine, W. V. (1953–1954). Oxford University lectures. Autograph manuscript. MS Am 2587: Box 116, Item 3277.
Quine, W. V. (1953–1954). Oxford University lecture: Philosophy of logic. Autograph manuscript. MS Am 2587: Box 118, Item 3283.

QUINE'S UNPROCESSED PAPERS

Quine's unprocessed archive is stored at Douglas B. Quine's house in Bethel, Connecticut. They will soon be added to the W. V. Quine Papers at Houghton Library. Unfortunately, there is no publicly available inventory of Quine's unprocessed archive yet. The references are ordered by box number.

Quine, W. V. (1924–1932). Correspondence between Quine and his parents. Box 21: Quine parents EasyClasp File: 1924–1932.
Quine, W. V. (1933–1965). Correspondence between Quine and his parents. Box 21: Quine parents EasyClasp File: 1933–1965.
Quine, W. V. (1933–1999). Datebooks. Box 45: Address Books, Datebooks, Diaries, Notepad.

QUINE'S LIBRARY

Quine's library is stored at the Harvard Depository and can be accessed at Houghton Library. A catalog of Quine's library can be found at http://hcl.harvard.edu/libraries/houghton/collections/modern/printed_acquisitions_1213.cfm. The references are ordered by call number.

Quine, W. V. (undated). Quine's copy of Dewey, J. (1925). *Experience and nature*. Item AC95.Qu441.Zz929d.
Quine, W. V. (undated). Quine's copy of Kuhn, T. (1962). *The structure of scientific Revolutions*. Item AC95.Qu441.Zz962k.
Quine, W. V. (undated). Quine's copy of Popper, K. (1962). *Conjectures and refutations*. Item AC95.Qu441.Zz962p.

PAPERS OF NELSON GOODMAN

Nelson Goodman's archive is stored at the Harvard University Archives. Unfortunately, there is no publicly available catalog of Goodman's papers.

Goodman, N. & Quine, W. V. (ca. 1935–1994). Correspondence. Papers of Nelson Goodman. Accession 14359, unlabeled box, unlabeled folder.

Published Literature

Almeder, R. (1998). *Harmless naturalism: The limits of science and the nature of philosophy*. Chicago: Open Court.
Alspector-Kelly, M. (2001). On Quine on Carnap on ontology. *Philosophical Studies, 102*(1), 93–122.
Ammerman, R. R. (1965). *Classics of analytic philosophy*. New York: McGraw-Hill.

Ariew, R. (1984). The Duhem thesis. *British Journal for the Philosophy of Science, 35*(4), 313–25.

Arnold, J. & Shapiro, S. (2007). Where in the (world wide) web of belief is the law of non-contradiction. *Noûs, 41*(2), 276–97.

Baldwin, T. (2007). Pragmatism and analysis. In M. Beaney (Ed.), *The analytic turn: Analysis in early analytic philosophy and phenomenology* (pp. 178–95). New York: Routledge.

Baldwin, T. (2013). C. I. Lewis and the analyticity debate. In Reck (2013a), pp. 201–27.

Balfour, A. J. (1895). *The foundations of belief. Being notes introductory to the study of theology.* New York: Longman, Green, and Co.

Barrett, R. B. & Gibson, R. F. (Eds.). (1990). *Perspectives on Quine.* Cambridge, MA: Basil Blackwell.

Beaney, M. (Ed.). (2013a). *The Oxford handbook of the history of analytic philosophy.* Oxford: Oxford University Press.

Beaney, M. (2013b). The historiography of analytic philosophy. In Beaney (2013a), pp. 30–60.

Becker, E. (2013). *The themes of Quine's philosophy: Meaning, reference, and knowledge.* New York: Cambridge University Press.

Becker, O. (1961). Review: *Word and object* by W. V. Quine. *Philosophische Rundschau, 9*(1), 238.

Ben-Menahem, Y. (2005). Black, white and grey: Quine on convention. *Synthese, 146*(3), 245–82.

Bennett, J. (1971). *Locke, Berkeley, Hume: Central themes.* Oxford: Oxford University Press.

Bergmann, G. (1950). A note on ontology. *Philosophical Studies, 1*(6), 89–92.

Berto, F. & Plebani, F. (2015). *Ontology and metaontology: A contemporary guide.* London: Bloomsbury.

Bird, G. (1995). Carnap and Quine: Internal and external questions. *Erkenntnis, 42*(1), 41–64.

Bourget, D. & Chalmers, D. (2014). What do philosophers believe? *Philosophical Studies, 170*(3), 465–500.

Brody, B. A. (1971). Review: *Words and objections* by D. Davidson & J. Hintikka (Eds.). *Studies in History and Philosophy of Science, Part A,* 2(2):167–75.

Büchner, L. F. (1855). *Kraft und Stoff. Empirisch-naturphilosophische Studien.* Frankfurt am Main: Verlag von Meidinger Sohn & Cie.

Burgess, J. (2004). Quine, analyticity and philosophy of mathematics. *Philosophical Quarterly, 54*(214), 38–55.

Carnap, R. (1928a). *Der Logische Aufbau der Welt.* Berlin-Schlachtensee: Weltkreis-Verlag. Second edition (1961). Hamburg: Felix Meiner Verlag. Translated by R. George (1967). *The logical structure of the world.* Berkeley: University of California Press.

Carnap, R. (1928b). *Scheinprobleme in der Philosophie: Das Fremdpsychische und der Realismusstreit.* Berlin-Schlachtensee: Weltkreis-Verlag. Translated by R. George (1967). *Pseudo-problems in philosophy: The heteropsychological and the realism controversy.* Berkeley: University of California Press.

Carnap, R. (1929). *Abriss der Logistik, mit Besonderer Berücksichtigung der Relationstheorie und ihrer Anwendungen.* Vienna: Verlag von Julius Springer.

Carnap, R. (1934). *Logische Syntax der Sprache.* Translated by A. Smeaton (1937). *The logical syntax of language.* London: Routledge & Kegan Paul, Trench, Trubner & Co.

Carnap, R. (1935). *Philosophy and logical syntax.* London: Kegan Paul, Trench, Trubner & Co.

Carnap, R. (1936). Testability and meaning. *Philosophy of Science, 3*(4), 419–71.

Carnap, R. (1937). Testability and meaning—continued. *Philosophy of Science, 4*(1), 1–40.

Carnap, R. (1942). *Introduction to semantics.* Cambridge, MA: Harvard University Press.

Carnap, R. (1947). *Meaning and necessity: A study in semantics and modal logic.* Enlarged edition (1956). Chicago: University of Chicago Press.

Carnap, R. (1950). Empiricism, semantics, and ontology. *Revue Internationale de Philosophie, 4*(11), 20–40. Reprinted in Carnap (1947/1956), pp. 205–21.

Carnap, R. (1954). Quine on analyticity. In Creath (1990), pp. 427–32.

Carnap, R. (1956). The methodological character of theoretical concepts. In H. Feigl & M. Scriven (Eds.), *Minnesota Studies in the Philosophy of Science,* volume 1 (pp. 38–76). Minneapolis: University of Minnesota Press.

Carnap, R. (1963a). Intellectual autobiography. In Schilpp (1963), pp. 3–84.

Carnap, R. (1963b). Replies and systematic expositions. In Schilpp (1963), pp. 859–1013.

Carnap, R. (1966). *Philosophical foundations of physics: An introduction to the philosophy of science.* Edited by M. Gardner. New York: Basic Books. Reprinted as *An introduction to the philosophy of science* (1974). New York: Basic Books.

Carnap, R. & Quine, W. V. (1932–1970). Correspondence. In Creath (1990), pp. 105–425.

Chomsky, N. (1959). Review: *Verbal Behavior* by B. F. Skinner. *Language, 35*(1), 26–58.

Chomsky, N. (1968). Quine's empirical assumptions. *Synthese, 19*(1), 53–68.

Coburn, R. C. (1964). Recent work in metaphysics. *American Philosophical Quarterly, 1*(3), 204–20.

Coffa, J. A. (1991). *The semantic tradition from Kant to Carnap.* Cambridge: Cambridge University Press.

Cohen, L. J. (1962). *The diversity of meaning.* London: Methuen.

Comte, A. (1830–1842). *Cours de Philosophie positive.* Paris: Bachelier. Translated and condensed by H. Martineau (1853). *The positive philosophy of Auguste Comte.* London: J. Chapman.

Creath, R. (1987). The initial reception of Carnap's doctrine of analyticity. *Noûs, 21*(4), 477–99.

Creath, R. (Ed). (1990). *Dear Carnap, Dear Van.* Berkeley: University of California Press.

Danto, A. C. (1967). Naturalism. In Edwards (1967), volume 6, pp. 448–50.

Davidson, D. & Hintikka, J. (1968). Editorial introduction. *Synthese, 19*(1–2), 1–2. Reprinted in Davidson & Hintikka (1969), pp. 1–2.

Davidson, D. & Hintikka, J. (Eds.). (1969). *Words and objections: Essays on the work of W. V. Quine.* Dordrecht: Reidel.

De Caro, M. & Macarthur, D. (Eds.). (2004a). *Naturalism in question.* Cambridge, MA: Harvard University Press.

De Caro, M. & Macarthur, D. (2004b). Introduction: The nature of naturalism. In De Caro & Macarthur (2004a), pp. 1–17.

De Caro, M. & Macarthur, D. (Eds.). (2010a). *Naturalism and normativity.* New York: Columbia University Press.

De Caro, M. & Macarthur, D. (2010b). Introduction: Science, naturalism, and the problem of normativity. In De Caro & Macarthur (2010a), pp. 1–19.

Decock, L. (2002a). *Trading ontology for ideology: The interplay of logic, set theory and semantics in Quine's philosophy.* Dordrecht: Kluwer Academic Publishers.

Decock, L. (2002b). A Lakatosian approach to the Quine-Maddy debate. *Logique et Analyse, 45*(179–80), 249–68.

Decock, L. (2004). Inception of Quine's ontology. *History and Philosophy of Logic, 25*(2), 111–29.

Decock, L. (2010). Quine's antimentalism in linguistics. *Logique et Analyse, 53*(212), 371–86.

Deledalle, G. (1962). Review: *Word and object* by W. V. Quine. *Les Études philosophiques, 17*(2), 278–79.

Dewey, J. (1925). *Experience and nature.* London: George Allen & Unwin (1929).

Dewey, J. (1944). Antinaturalism in extremis. In Krikorian (1944), pp. 1–16.

Dewey, J., Hook, S. & Nagel, E. (1945). Are naturalists materialists? *Journal of Philosophy, 42*(19), 515–30.

Dieterle, J. M. (1999). Mathematical, astrological, and theological naturalism. *Philosophia Mathematica, 7*(2), 129–35.

Dreben, B. (1990). Quine. In Barrett & Gibson (1990), pp. 81–85.

Dreben, B. (1992). Putnam, Quine—and the facts. *Philosophical Topics, 20*(1), 293–315.

Duhem, P. (1914). *La théorie physique: Son objet, sa structure.* Paris: Marcel Rivière & Cie. Translated by P. P. Weiner (1954). *The aim and structure of physical theory.* Renewed edition (1982). Princeton, NJ: Princeton University Press.

Dupré, J. (2004). The miracle of monism. In De Caro & Macarthur (2004a), pp. 36–58.

Ebbs, G. (2011a). Carnap and Quine on truth by convention. *Mind, 120*(478), 193–237. Reprinted in Ebbs (2017), pp. 57–94.

Ebbs, G. (2011b). Quine gets the last word. *Journal of Philosophy, 108*(11), 617–32. Reprinted in Ebbs (2017), pp. 113–27.

Ebbs, G. (2016). Quine's "predilection" for finitism. *Metascience, 25*(1), 31–36.

Ebbs, G. (2017). *Carnap, Quine, and Putnam on methods of inquiry.* Cambridge: Cambridge University Press.

Edwards, P. (Ed.). (1967). *The encyclopedia of philosophy.* New York: Macmillan.

Einstein, A. (1935). *The world as I see it*. London: John Lane The Bodley Head Limited.

Eklund, M. (2013). Carnap's metaontology. *Noûs, 47*(2), 229–49.

Eldridge, M. (2004). Naturalism. In A. T. Marsoobian & J. Ryder (Eds.), *The Blackwell guide to American philosophy* (pp. 52–71). Malden, MA: Blackwell Publishing.

Floyd, J. (2009). Recent themes in the history of early analytic philosophy. *Journal of the History of Philosophy, 47*(2), 157–200.

Floyd, J. & Shieh, S. (Eds.). (2001). *Future pasts: The analytic tradition in twentieth-century philosophy*. New York: Oxford University Press.

Fogelin, R. J. (2004). Aspects of Quine's naturalized epistemology. In Gibson (2004), pp. 19–46.

Føllesdal, D. (2001). Quine—the philosophers' philosopher. *Dialectica, 55*(2), 99–103.

Føllesdal, D. (2011). Developments in Quine's behaviorism. *American Philosophical Quarterly, 48*(3), 273–82.

Frege, G. (1884). *Die Grundlagen der Arithmetik: Eine logisch mathematische Untersuchung über den Begriff der Zahl*. Breslau: Wilhelm Koebner. Translated by D. Jacquette (2007). *The foundations of arithmetic: A logical-mathematical investigation into the concept of number*. New York: Pearson Longman.

Friedman, M. (1987). Carnap's *Aufbau* reconsidered. *Noûs, 21*(4), 521–45.

Friedman, M. (1992). Epistemology in the *Aufbau*. *Synthese, 93*(1–2), 15–57.

Friedman, M. (2007). Introduction: Carnap's revolution in philosophy. In Friedman & Creath (2007), pp. 1–18.

Friedman, M. & Creath, R. (Eds.). (2007). *The Cambridge companion to Carnap*. Cambridge: Cambridge University Press.

Frost-Arnold, G. (2011). Quine's evolution from "Carnap's disciple" to the author of "Two dogmas." *HOPOS: Journal of the International Society for the History of Philosophy of Science, 1*(2), 291–316.

Frost-Arnold, G. (2013). *Carnap, Tarski, and Quine at Harvard: Conversations on logic, mathematics, and science*. Chicago: Open Court.

Galanter, E. H. (1961). Analysis of language with a behavioral bias. Review: *Word and object* by W. V. Quine. *Contemporary Psychology, 6*(12), 430–31.

Gallois, A. (1998). Does ontology rest on a mistake? *Proceedings of the Aristotelian Society, Supplementary Volume, 72*(1), 263–83.

Gardiner, A. H. (1940). *The theory of proper names: A controversial essay*. Oxford: Oxford University Press.

Gaudet, E. (2006). *Quine on meaning: The indeterminacy of translation*. London: Continuum.

Geach, P. T. (1961). Review: *Word and object* by W. V. Quine. *Philosophical Books, 2*(1), 14–17.

Giedymin, J. (1972). Quine's philosophical naturalism. *British Journal for the Philosophy of Science, 23*(1), 45–55.

Gibson, R. F. (1982). *The philosophy of W. V. Quine: An expository essay*. Tampa: University Presses of Florida.

Gibson, R. F. (1988). *Enlightened empiricism: An examination of W. V. Quine's theory of knowledge*. Tampa: University of South Florida Press.

Gibson, R. F. (1992). The key to interpreting Quine. *Southern Journal of Philosophy, 30*(4), 17–30.

Gibson, R. F. (Ed.). (2004). *The Cambridge companion to Quine*. Cambridge: Cambridge University Press.

Glock, H. J. (2002). Does ontology exist? *Philosophy, 77*(2), 235–60.

Glock, H. J. (2003). *Quine and Davidson on language, thought and reality*. New York: Cambridge University Press.

Glock, H. J. (2008a). *What is analytic philosophy?* New York: Cambridge University Press.

Glock, H. J. (2008b). Analytic philosophy and history: A mismatch? *Mind, 117*(468), 867–97.

Godfrey-Smith, P. (2014). Quine and pragmatism. In Harman & Lepore (2014), pp. 54–68.

Goldfarb, W. (2008). On Quine's philosophy. Centennial celebration and marker dedication honoring W. V. Quine at Oberlin College, June 25, 2008. Retrieved from http://www.wvquine.org/GoldfarbOberlin.pdf.

Goodman, N. & Quine, W. V. (1947). Steps toward a constructive nominalism. *Journal of Symbolic Logic, 12*(4), 105–22.

Goodman, N., Quine, W. V., & White, M. (1947). Appendix: Nelson Goodman, W. V. Quine, and Morton White: A triangular correspondence in 1947. In White (1999), pp. 337–57.

Goodstein, R. L. (1961). Review: *Word and object* by W. V. Quine. *Philosophy of Science, 28*(2), 217.

Gosselin, M. (1990). *Nominalism and contemporary nominalism: Ontological and epistemological implications of the work of W. V. O. Quine and of N. Goodman*. Dordrecht: Springer.

Grandy, R. E. (1973). Review: *Words and objections* by D. Davidson & J. Hintikka (Eds). *Philosophical Review, 82*(1), 99–110.

Greenlee, D. (1973). Relativity without inscrutability. *Philosophy and Phenomenological Research, 33*(4), 574–78.

Gregory, P. A. (2008). *Quine's naturalism: Language, theory and the knowing subject*. London: Continuum.

Grice, H. P. & Strawson, P. F. (1956). In defense of a dogma. *Philosophical Review, 65*(2), 141–58.

Grize, J. (1965). Genetic epistemology and psychology. In B. B. Wolman & E. Nagel (Eds.), *Scientific psychology: Principles and approaches* (pp. 460–73). New York: Basic Books.

Guttenplan, S. (Ed.). (1975). *Mind and language*. Oxford: Oxford University Press.

Haack, S. (1976). Some preliminaries to ontology. *Journal of Philosophical Logic, 5*(4), 457–74.

Haack, S. (1977). Analyticity and logical truth in *Roots of reference*. *Theoria, 43*(2), 129–43.

Haack, S. (1993a). *Evidence and inquiry*. Second and expanded edition (2009). Amherst: Prometheus Books.

Haack, S. (1993b). The two faces of Quine's naturalism. *Synthese, 94*(3), 335–56.

Hacker, P. M. S. (1996). *Wittgenstein's place in twentieth-century analytic philosophy*. Oxford: Blackwell.

Hacker, P. M. S. (1997). The rise of twentieth century analytic philosophy. In H. J. Glock (Ed.), *The rise of analytic philosophy* (pp. 51–76). Oxford: Blackwell.

Hacker, P. M. S. (2006). Passing by the naturalistic turn: On Quine's cul-de-sac. *Philosophy, 81*(316), 231–53.

Haeckel, E. (1899). *Die welträthsel: Gemeinverständliche Studien über monistische Philosophie*. Bonn: Verlag von Emil Strauss.

Hahn, L. & Schilpp, P. (Eds.). (1986). *The philosophy of W. V. Quine*. Library of Living Philosophers. Volume 18. Expanded edition (1998). La Salle, IL: Open Court.

Hale, B. (1999). Review: *Naturalism in mathematics* by P. Maddy. *Journal of Symbolic Logic, 64*(1), 394–96.

Haller, R. (1992). Alfred Tarski: Drei Briefe an Otto Neurath. *Grazer Philosophische Studien, 43*(1), 1–32.

Harding, S. (Ed). (1976). *Can theories be refuted? Essays on the Duhem-Quine thesis*. Dordrecht: Reidel.

Hare, P. H. (1968). Purposes and methods of writing the history of recent American philosophy. *Southern Journal of Philosophy, 6*(4), 269–78.

Harman, G. (1967). Quine on meaning and existence, I. The death of meaning. *Review of Metaphysics, 21*(1), 124–51.

Harman, G. & Lepore, E. (Eds.). (2014). *A companion to W. V. O. Quine*. Chichester: Wiley Blackwell.

Hempel, C. (1952). *Fundamentals of concept formation in the empirical sciences. International encyclopedia of unified science*, volume 2, number 7. Chicago: University of Chicago Press.

Hickman, L. A. (Ed.). (2008). *The correspondence of John Dewey*. Volume 4: *1953–2007*. Third edition. Electronic edition. Charlottesville, VA: InteLex Corporation.

Hockney, D. (1973). Conceptual structures. In G. Pearce & P. Maynard (Eds.), *Conceptual change* (pp. 141–66). Dordrecht: Reidel.

Honderich, T. (Ed.). (1995). *The Oxford companion to philosophy*. Second edition (2005). Oxford: Oxford University Press.

Hornsby, J. (1997). *Simple mindedness: In defense of naïve naturalism in the philosophy of mind*. Cambridge, MA: Harvard University Press.

Hookway, C. (1988). *Quine: Language, experience and reality*. Cambridge: Polity Press.

Huxley, T. (1892). *Essays upon some controverted questions*. London: Macmillan.

Hylton, P. (1990). *Russell, idealism, and the emergence of analytic philosophy*. Oxford: Clarendon Press.

Hylton, P. (2001). "The defensible province of philosophy": Quine's 1934 lectures on Carnap. In Floyd & Shieh (2001), pp. 257–76.
Hylton, P. (2002). Analyticity and holism in Quine's thought. *Harvard Review of Philosophy, 10*(1), 11–26.
Hylton, P. (2007). *Quine*. New York: Routledge.
Hylton, P. (2014a). Quine's naturalism revisited. In Harman & Lepore (2014), pp. 148–62.
Hylton, P. (2014b). Significance in Quine. *Grazer Philosophische Studien, 89*(1), 113–33.
Isaac, J. (2005). W. V. Quine and the origins of analytic philosophy in the United States. *Modern Intellectual History, 2*(2), 205–34.
Isaac, J. (2011). Missing links: W. V. Quine, the making of "Two dogmas," and the analytic roots of post-analytic philosophy. *History of European Ideas, 37*(3), 267–79.
Janssen-Lauret, F. & Kemp, G. (Eds.). (2016). *Quine and his place in history*. Houndmills: Palgrave Macmillan.
Jewett, A. (2011). Canonizing Dewey: Naturalism, logical empiricism, and the idea of American philosophy. *Modern Intellectual History, 8*(1), 91–125.
Johnsen, B. (2005). How to read "Epistemology naturalized." *Journal of Philosophy, 102*(2), 78–93.
Johnstone, H. W. (1961). Review: *Word and object* by W. V. Quine. *Philosophy and Phenomenological Research, 22*(1), 115–16.
Joll, N. (2010). Contemporary metaphilosophy. *Internet Encyclopedia of Philosophy*. Retrieved from http://www.iep.utm.edu/con-meta/.
Katz, J. (1990). The refutation of indeterminacy. In Barrett & Gibson (1990), pp. 177–97.
Kemp, G. (2006). *Quine: A guide for the perplexed*. New York: Continuum.
Kemp, G. (2012). *Quine vs. Davidson: Truth, reference, and meaning*. Oxford: Oxford University Press.
Kertész, A. (2002). On the de-naturalization of epistemology. *Journal for the General Philosophy of Science, 33*(2), 269–88.
Kim, J. (1988). What is "naturalized epistemology?" *Philosophical Perspectives, 2*(1), 381–405.
Kim, J. (2003). The American origins of philosophical naturalism. *Journal of Philosophical Research*, APA Centennial Supplement: Philosophy in America at the turn of the century, *28*(s), 83–98.
Kitcher, P. (1992). The naturalists return. *Philosophical Review, 101*(1), 53–114.
Klein, A. (2017). Russell on acquaintance with spatial properties: The significance of James. In C. Pincock & S. Lapointe (Eds.), *Innovations in the History of Analytical Philosophy* (pp. 229–64). Houndmills: Palgrave Macmillan.
Kmita, J. (1971). Meaning and functional reason. *Quality and Quantity, 5*(2), 353–68.
Koppelberg, D. (1987). *Die Aufhebung der analytischen Philosophie: Quine als Synthese von Carnap und Neurath*. Frankfurt am Main: Suhrkamp Verlag.
Koskinen, H. & Pihlström, S. (2006). Quine and pragmatism. *Transactions of the Charles S. Peirce Society*, 42(3), 309–46.
Krikorian, Y. (Ed.). (1944). *Naturalism and the human spirit*. New York: Columbia University Press.
Kripke, S. (1982). *Wittgenstein on rules and private language*. Oxford: Blackwell.
Kuhn, T. (1962). *The structure of scientific revolutions*. Chicago: University of Chicago Press.
Kusch, M. (1995). *Psychologism. A case study in the sociology of philosophical knowledge*. London: Routledge.
Lakatos, I. & Musgrave, A. (Eds.). (1968). *Problems in the philosophy of science*. Amsterdam: North-Holland.
Larrabee, H. A. (1944). Naturalism in America. In Krikorian (1944), pp. 319–53.
Leiter, B. (2004). Introduction. In Leiter (Ed.), *The future to philosophy* (pp. 1–23). Oxford: Oxford University Press.
Leonardi, P. & Santambrogio, M. (Eds.). (1995). *On Quine: New essays*. New York: Cambridge University Press.
Lewes, G. H. (1874). *Problems of life and mind*. London: Trübner.
Lewis, C. I. (1929). *Mind and the world-order: Outline of a theory of knowledge*. New York: Charles Scribner's Sons.
Lewis, C. I. (1946). *An analysis of knowledge and valuation*. La Salle, IL: Open Court.

Lieb, I. C. (1962). Review: *Word and object* by W. V. Quine. *International Philosophical Quarterly, 2*(1), 92–109.

Loeffler, R. (2005). Intertheoretical identity and ontological reductions. *Erkenntnis, 62*(2), 157–87.

Loux, M. J. (1972). Recent work in ontology. *American Philosophical Quarterly, 9*(2), 119–38.

Lugg, A. (2012). W. V. Quine on analyticity: "Two dogmas of empiricism" in context. *Dialogue, 51*(2), 231–46.

Lugg, A. (2016). Quine, Wittgenstein and "The abyss of the transcendental." In Janssen-Lauret & Kemp (2016), pp. 198–209.

Macarthur, D. (2008). Quinean naturalism in question. *Philo, 11*(1), 1–14.

MacIntyre, A. (1984). The relationship of philosophy to its past. In Rorty et al. (1984), pp. 31–48.

Maddy, P. (1997). *Naturalism in mathematics*. Oxford: Oxford University Press.

Maddy, P. (2005). Three forms of naturalism. In Shapiro (2005), pp. 437–59.

Maddy, P. (2007). *Second philosophy: A naturalistic method*. Oxford: Oxford University Press.

Malcolm, N. (1940). Are necessary propositions really verbal? *Mind, 49*(194), 189–203.

Mancosu, P. (2005). Harvard 1940–1941: Tarski, Carnap and Quine on a finitistic language of mathematics for science. *History and Philosophy of Logic, 26*(4), 327–57.

Mancosu, P. (2008). Quine and Tarski on nominalism. In Zimmerman (2008), pp. 22–55.

Maxwell, G. (1968). Scientific methodology. In Lakatos & Musgrave (1968), pp. 148–60.

Margolis, J. (1969). Some ontological policies. *Monist, 53*(2), 231–45.

Massey, G. J. (2011). Quine and Duhem on holistic hypothesis testing. *American Philosophical Quarterly, 48*(3), 239–66.

Mill, J. S. (1865). *Examination of Sir William Hamilton's philosophy*. London: Longman, Green, and Co.

Moore, G. E. (1903). *Principia ethica*. Cambridge: Cambridge University Press.

Morris, S. (2015). Quine, Russell, and naturalism: From a logical point of view. *Journal of the History of Philosophy, 53*(1), 133–55.

Morris, S. (forthcoming). Quine against Lewis (and Carnap) on truth by convention. *Pacific Philosophical Quarterly*. http://doi.org/10.1111/papq.12185.

Morrison, J. (2010). Just how controversial is evidential holism? *Synthese, 173*(3), 335–52.

Moser, P. & Yandell, D. (2000). Farewell to philosophical naturalism. In W. Craig & J. Moreland (Eds.), *Naturalism: A critical analysis* (pp. 3–23). London: Routledge.

Murphey, M. (2012). *The development of Quine's philosophy*. Dordrecht: Springer.

Næss, A. (1938). *"Truth" as conceived by those who are not professional philosophers*. Oslo: I. Kommisjon Hos Jacob Dybwad.

Nagel, E. (1954). Naturalism reconsidered. *Proceedings and Addresses of the American Philosophical Association, 28*(1):5–17.

Nagel, T. (1986). *The view from nowhere*. New York: Oxford University Press.

Nasim, O. W. (2012). The spaces of knowledge: Bertrand Russell, logical construction, and the classification of the sciences. *British Journal for the History of Philosophy, 20*(6), 1163–82.

Oesterle, J. A. (1961). Review: *Word and object* by W. V. Quine. *Thomist, 24*(1), 117–20.

Papineau, D. (2009). Naturalism. In E. Zalta (Ed.), *The Stanford Encyclopedia of Philosophy* (Spring 2009 Edition). Retrieved from http://plato.stanford.edu/archives/spr2009/entries/naturalism/.

Parent, T. (2008). Quine and logical truth. *Erkenntnis, 68*(1), 103–12.

Paseau, A. (2013). Naturalism in the philosophy of mathematics. In E. Zalta (Ed.), *The Stanford encyclopedia of philosophy* (Spring 2013 edition). Retrieved from http://plato.stanford.edu/archives/sum2013/entries/naturalism-mathematics/.

Passmore, J. (1966). *A hundred years of philosophy*. Second edition. London: Duckworth.

Peijnenburg, J. (2000). Identity and difference: A hundred years of analytic philosophy. *Metaphilosophy, 31*(4), 365–81.

Pincock, C. (2007). Carnap, Russell, and the external world. In Friedman & Creath (2007), pp. 106–28.

Popper, K. R. (1962). *Conjectures and refutations: The growth of scientific knowledge*. New York: Basic Books.

Presley, C. F. (1961). Review: *Word and object* by W. V. Quine. *Australasian Journal of Philosophy, 39*(2), 175–90.

Presley, C. F. (1967). Quine, Willard Van Orman. In Edwards (1967), volume 7, pp. 53–55.

Price, H. (2007). Quining naturalism. *Journal of Philosophy, 104*(8), 375–402.

Price, H. (2009). Metaphysics after Carnap: The ghost who walks? In D. Chalmers, D. Manley & R. Wasserman (Eds.), *Metametaphysics: New essays on the foundations of ontology* (pp. 320–46). Oxford: Oxford University Press.

Putnam, H. (1982). Why reason can't be naturalized. *Synthese, 52*(1), 3–24. Reprinted in Putnam (1983), pp. 229–47.

Putnam, H. (1983). *Realism and reason. Philosophical Papers.* Volume 3. New York: Cambridge University Press.

Putnam, H. (2002). "Quine." *Common Knowledge, 8*(2), 273–79.

Putnam, H. (2004). *Ethics without ontology.* Cambridge, MA: Harvard University Press.

Quine, W. V. (1934a). *A system of logistic.* Cambridge, MA: Harvard University Press.

Quine, W. V. (1934b). Lectures on Carnap. In Creath (1990), pp. 45–103.

Quine, W. V. (1934c). Ontological remarks on the propositional calculus. *Mind, 43*(172), 472–76. Reprinted in Quine (1966a), pp. 265–71.

Quine, W. V. (1936). Truth by convention. In O. Lee (Ed.), *Philosophical essays for A. N. Whitehead* (pp. 90–124). New York: Longmans. Reprinted in Quine (1966a), pp. 77–106.

Quine, W. V. (1937). On Cantor's theorem. *Journal of Symbolic Logic, 2*(3), 120–24.

Quine, W. V. (1939a). A logistical approach to the ontological problem. In Quine (1966a), pp. 197–202.

Quine, W. V. (1939b). Designation and existence. *Journal of Philosophy, 36*(26), 701–9.

Quine, W. V. (1940). *Mathematical logic.* New York: Norton.

Quine, W. V. (1941a). *Elementary logic.* Boston: Ginn.

Quine, W. V. (1941b). Whitehead and the rise of modern logic. In P. A. Schilpp (Ed.), *The philosophy of Alfred North Whitehead* (pp. 125–63). Evanston, IL: Northwestern University Press. Reprinted in Quine (1966b), pp. 3–36.

Quine, W. V. (1943a). *O sentido da nova lógica.* São Paulo: Livraria Martins (1944).

Quine, W. V. (1943b). Notes on existence and necessity. *Journal of Philosophy,* 40(5):113–27.

Quine, W. V. (1946a). Nominalism. Presented at the Harvard philosophical colloquium in Emerson B, Cambridge, MA, March 11, 1946. Transcribed by P. Mancosu. In Zimmerman (2008), pp. 3–21. Reprinted in Quine (2008a), pp. 9–23.

Quine, W. V. (1946b). On the notion of an analytic statement. Presented at the University of Pennsylvania, Philadelphia, December 18, 1946. Transcribed and edited by G. Zanet. In Quine (2008a), pp. 24–35.

Quine, W. V. (1946c). Lectures on David Hume's philosophy. Delivered at Harvard University, summer 1946. Edited by J. Buickerood. In Quine (2008a), pp. 36–136.

Quine, W. V. (1947). Review: Contradiction and the presupposition of existence, by E. J. Nelson. *Journal of Symbolic Logic, 12*(2), 52–55.

Quine, W. V. (1948). On what there is. *Review of Metaphysics, 2*(5), 21–38. Reprinted in Quine (1953a), pp. 1–19.

Quine, W. V. (1949). Animadversions on the notion of meaning. Presented at the Philosophy Colloquium, University of Pennsylvania, Philadelphia, December 6, 1949. Transcribed by G. Zanet & A. Quine. In Quine (2008a), pp. 152–56.

Quine, W. V. (1950a). *Methods of logic.* First edition. New York: Henry Holt.

Quine, W. V. (1950b). Identity, ostension, hypostasis. *Journal of Philosophy, 47*(22), 621–33. Reprinted in Quine (1953a), pp. 65–79.

Quine, W. V. (1951a). Two dogmas of empiricism. *Philosophical Review, 60*(1), 20–43. Reprinted in Quine (1953a), pp. 20–46.

Quine, W. V. (1951b). Ontology and ideology. *Philosophical Studies, 2*(1), 11–15.

Quine, W. V. (1951c). On Carnap's views on ontology. *Philosophical Studies, 2*(5), 65–72. Reprinted in Quine (1966a), pp. 197–202.

Quine, W. V. (1952). On mental entities. *Proceedings of the American Academy of Arts and Sciences* (1953), *80*(3), 198–203. Reprinted in Quine (1966a), pp. 221–27.

Quine, W. V. (1953a). *From a logical point of view.* Second edition (1961). Cambridge, MA: Harvard University Press.

Quine, W. V. (1953b). Mr. Strawson on logical theory. *Mind, 62*(248), 433–51. Reprinted in Quine (1966a), pp. 137–57.

Quine, W. V. (1954a). Carnap and logical truth. *Synthese* (1960), *12*(4), 350–74. Reprinted in Quine (1966a), pp. 107–32.

Quine, W. V. (1954b). The scope and language of science. In L. Leary (Ed.) (1955), *The unity of knowledge* (pp. 231–47). New York: Doubleday. Reprinted in Quine (1966a), pp. 228–45.

Quine, W. V. (1955). Posits and reality. In S. Uyeda (Ed.). (1960), *Basis of the contemporary philosophy.* Volume 5 (pp. 391–400). Tokyo: Waseda University Press. Reprinted in Quine (1966a), pp. 246–54.

Quine, W. V. (1960). *Word and object.* Cambridge, MA: MIT Press.

Quine, W. V. (1962). A comment on Grünbaum's claim. Letter to Adolf Grünbaum, June 1, 1962. In Harding (1976), p. 132.

Quine, W. V. (1963). *Set theory and its logic.* Revised edition (1969). Cambridge, MA: Belknap Press.

Quine, W. V. (1966a). *The ways of paradox and other essays.* New York: Random House. Revised edition (1976). Cambridge, MA: Harvard University Press.

Quine, W. V. (1966b). *Selected logic papers.* New York: Random House. Enlarged edition (1995). Cambridge, MA: Harvard University Press.

Quine, W. V. (1966c). Russell's ontological development. *Journal of Philosophy, 63*(21), 657–67. Reprinted in Quine (1981a), pp. 73–85.

Quine, W. V. (1968a). Ontological relativity. *Journal of Philosophy, 65*(7), 185–212. Reprinted in Quine (1969a), pp. 26–68.

Quine, W. V. (1968b). Existence and quantification. *L'Age de la Science, 1*(1), 151–64. Reprinted in Quine (1969a), pp. 91–113.

Quine, W. V. (1968c). Responding to Grover Maxwell. In Lakatos & Musgrave (1968), pp. 161–63. Reprinted in Quine (1981a), pp. 176–78.

Quine, W. V. (1968d). Reply to Smart. *Synthese, 19*(1–2), 264–66. Reprinted in Davidson & Hintikka (1969), pp. 292–94.

Quine, W. V. (1968e). Reply to Stenius. *Synthese, 19*(1–2), 270–73. Reprinted in Davidson & Hintikka (1969), pp. 298–301.

Quine, W. V. (1968f). Reply to Chomsky. *Synthese, 19*(1–2), 274–83. Reprinted in Davidson & Hintikka (1969), pp. 302–11.

Quine, W. V. (1968g). Reply to Davidson. *Synthese, 19*(1–2), 303–5. Reprinted in Davidson & Hintikka (1969), pp. 333–35.

Quine, W. V. (1968h). Reply to Stroud. *Synthese, 19*(1–2), 288–91. Reprinted in Davidson & Hintikka (1969), pp. 316–18.

Quine, W. V. (1968i). Linguistics and philosophy. In S. Hook (Ed.) (1969), *Language and philosophy: A symposium* (pp. 95–98). New York University Institute of Philosophy, 9th, 1968. New York: New York University Press. Reprinted in Quine (1966a), pp. 56–58.

Quine, W. V. (1969a). *Ontological relativity and other essays.* John Dewey Essays in Philosophy. New York: Columbia University Press.

Quine, W. V. (1969b). Epistemology naturalized. In *Proceedings of the XIVth International Congress of Philosophy, Vienna, 2nd to 9th September 1968,* pp. 87–103. Vienna: Herder (1971). Reprinted in Quine (1969a), pp. 69–90.

Quine, W. V. (1969c). Natural kinds. In N. Rescher (Ed.) (1970), *Essays in honor of Carl G. Hempel: A tribute on the occasion of his sixty-fifth birthday* (pp. 5–23). Reprinted in Quine (1969a), pp. 114–38.

Quine, W. V. (1970a). *Philosophy of logic.* Englewood Cliffs, NJ: Prentice-Hall. Second edition (1986). Cambridge, MA: Harvard University Press.

Quine, W. V. (1970b). Philosophical progress in language theory. *Metaphilosophy, 1*(1), 2–19.

Quine, W. V. (1970c). Grades of theoreticity. In L. Foster & J. Swanson (Eds.), *Experience & theory* (pp. 1–17). Amherst: University of Massachusetts Press.

Quine, W. V. (1972). Vagaries of definition. *Annals of the New York Academy of Sciences,* 211(1): 247–50 (1973). Reprinted in Quine (1966a), pp. 50–55.

Quine, W. V. (1973). *The roots of reference.* Paul Carus Lectures. La Salle, IL: Open Court. (1974).

Quine, W. V. (1975a). Five milestones of empiricism. In Quine (1981a), pp. 67–72.
Quine, W. V. (1975b). The pragmatists' place in empiricism. In R. Mulvaney & P. Zeltner (Eds.) (1981), *Pragmatism: Its sources and prospects* (pp. 21–39). Columbia: University of South Carolina Press.
Quine, W. V. (1975c). On empirically equivalent systems of the world. *Erkenntnis, 9*(3), 313–28. Reprinted in Quine (2008a), pp. 228–43.
Quine, W. V. (1975d). The nature of natural knowledge. In Guttenplan (1975), pp. 67–81. Reprinted in Quine (2008a), pp. 257–70.
Quine, W. V. (1977). Facts of the matter. In R. W. Shahan & K. R. Merrill (Eds.), *American philosophy from Edwards to Quine* (pp. 176–96). Norman: University of Oklahoma Press. Reprinted in Quine (2008a), pp. 271–86.
Quine, W. V. (1978a). The ideas of Quine. Interview by B. Magee. In B. Magee (Ed.), *Men of ideas* (pp. 168–79). London: BBC Publications. Reprinted in Quine (2008b), pp. 5–17.
Quine, W. V. (1978b). Otherworldly. *New York Review of Books*, November 23, 25(18):25. Reprinted as "Goodman's *Ways of worldmaking*" in Quine (1981a), pp. 96–99.
Quine, W. V. (1981a). *Theories and things*. Cambridge, MA: Harvard University Press.
Quine, W. V. (1981b). Things and their place in theories. In Quine (1981a), pp. 1–23.
Quine, W. V. (1981c). What price bivalence? *Journal of Philosophy, 78*(2), 90–95. Reprinted in Quine (1981a), pp. 31–37.
Quine, W. V. (1981d). On the very idea of a third dogma. In Quine (1981a), pp. 38–42.
Quine, W. V. (1981e). Reply to Stroud. *Midwest Studies in Philosophy, 6*(1), 473–75.
Quine, W. V. (1982). *Methods of logic*. Fourth edition, revised and enlarged. Cambridge, MA: Harvard University Press.
Quine, W. V. (1984a). Carnap's positivistic travail. *Fundamenta Scientiae, 5*(1), 325–33. Reprinted in Quine (2008b), pp. 119–28.
Quine, W. V. (1984d). What I believe. In M. Booth (Ed.), *What I believe: 13 eminent people of our time argue for their philosophy of life* (pp. 69–75). New York: Waterstone / Firethorn Press. Reprinted in Quine (2008b), pp. 307–11.
Quine, W. V. (1985a). *The time of my life: An autobiography*. Cambridge, MA: MIT Press.
Quine, W. V. (1985b). Four hot questions in philosophy. Review of Strawson (1983). *New York Review of Books*, February 14, 32(2), 32. Reprinted in Quine (2008b), pp. 206–10.
Quine, W. V. (1986a). Autobiography of W. V. Quine. In Hahn & Schilpp (1986), pp. 1–46.
Quine, W. V. (1986b). Reply to Geoffrey Hellmann. In Hahn & Schilpp (1986), pp. 206–8.
Quine, W. V. (1986c). Reply to Robert Nozick. In Hahn & Schilpp (1986), pp. 364–67.
Quine, W. V. (1986d). Reply to Charles Parsons. In Hahn & Schilpp (1986), pp. 396–403.
Quine, W. V. (1986e). Reply to Hilary Putnam. In Hahn & Schilpp (1986), pp. 427–31.
Quine, W. V. (1986f). Reply to Jules Vuillemin. In Hahn & Schilpp (1986), pp. 619–22.
Quine, W. V. (1986g). Reply to Morton White. In Hahn & Schilpp (1986), pp. 663–65.
Quine, W. V. (1986h). The way the world is. Presented at Harvard's 350th celebration, March 1986. In Quine (2008a), pp. 166–71.
Quine, W. V. (1986i). The sensory support of science. In *Discursos pronunciados en el acto de investidura Doctor "Honoris Causa" de los Profesores Diego Angulo Iniguez, W. V. Quine, Arthur M. Silverstein, J.D. Smyth*, pp. 29–60. Granada: Universidad de Granada. Reprinted in Quine (2008a), pp. 327–37.
Quine, W. V. (1987a). Carnap. *Yale Review, 76*(2), 226–30. Reprinted in Quine (2008b), pp. 142–45.
Quine, W. V. (1987b). Indeterminacy of translation again. *Journal of Philosophy, 84*(1), 5–10. Reprinted in Quine (2008a), pp. 341–46.
Quine, W. V. (1988). Quine speaks his mind. Interview by E. Pivcevic. *Cogito, 2*(2), 1–5. Reprinted in Quine (2008b), pp. 21–29.
Quine, W. V. (1990a). *Pursuit of truth.* Revised edition (1992). Cambridge, MA: Harvard University Press.
Quine, W. V. (1990b). Three indeterminacies. In Barrett & Gibson (1990), pp. 1–16.
Quine, W. V. (1990c). Comment on Berger. In Barrett & Gibson (1990), pp. 36–37.
Quine, W. V. (1991a). Immanence and validity. *Dialectica, 45*(2–3), 219–30. Reprinted in Quine (1966b), pp. 242–50.

Quine, W. V. (1991b). Two dogmas in retrospect. *Canadian Journal of Philosophy, 21*(3), 265–74. Reprinted in Quine (2008a), pp. 390–400.

Quine, W. V. (1992). Structure and nature. *Journal of Philosophy, 89*(1), 5–9. Reprinted in Quine (2008a), pp. 401–6.

Quine, W. V. (1994a). Truth. In G. Fløistad (Ed.), *Philosophical problems today*. Volume 1 (pp. 1–20). Dordrecht: Kluwer Academic Publishers. Reprinted in Quine (2008a), pp. 420–37.

Quine, W. V. (1994a). Comment on Neil Tennant's "Carnap and Quine." In Salmon & Wolters (1994), pp. 345–51. Reprinted in Quine (2008b), pp. 216–22.

Quine, W. V. (1994b). Interview with Willard Van Orman Quine. Interview by L. Bergström & D. Føllesdal. *Theoria, 60*(3), 193–206. Reprinted in Quine (2008b), pp. 69–81.

Quine, W. V. (1994c). Exchange between Donald Davidson and W. V. Quine following Davidson's lecture. *Theoria, 60*(3), 226–31. Reprinted in Quine (2008b), pp. 152–56.

Quine, W. V. (1994d). Responses to articles by Abel, Bergström, Davidson, Dreben, Gibson, Hookway, and Prawitz. *Inquiry, 37*(4), 495–505. Reprinted in Quine (2008b), pp. 223–34.

Quine, W. V. (1995a). *From stimulus to science*. Cambridge, MA: Harvard University Press.

Quine, W. V. (1995b). Contextual definition. In Honderich (1995), p. 169.

Quine, W. V. (1995c). Naturalism; Or, living within one's means. *Dialectica, 49*(2–4), 251–61. Reprinted in Quine (2008a), pp. 461–72.

Quine, W. V. (1995d). Reactions. In Leonardi & Santambrogio (1995), pp. 347–61. Reprinted in Quine (2008b), pp. 235–50.

Quine, W. V. (1995e). Response to Gary Ebbs. In Janssen-Lauret & Kemp (2016), pp. 33–36.

Quine, W. V. (1996a). Progress on two fronts. *Journal of Philosophy, 93*(4), 159–63. Reprinted in Quine (2008a), pp. 473–77.

Quine, W. V. (1996b). Quine /'kwain/, Willard Van Orman (b. 1908). In T. Mautner (Ed.), *Dictionary of philosophy*. Oxford: Blackwell Publishers. Reprinted in Quine (2008b), pp. 346–48.

Quine, W. V. (1997). Responses to essays by Smart, Orenstein, Lewis and Holdcroft, and Haack. *Revue Internationale de Philosophie, 51*(4), 567–82. Partly reprinted in Quine (2008b), pp. 251–56.

Quine, W. V. (1999a). Where do we disagree. In L. Hahn (Ed.), *Philosophy of Donald Davidson*. Library of Living Philosophers. Volume 27 (pp. 73–79). Chicago: Open Court. Reprinted in Quine (2008b), pp. 159–65.

Quine, W. V. (1999b). Responses to Szubka, Lehrer, Bergström, Gibson, Miscevic, and Orenstein. In A. Orenstein & P. Kotatko (Eds.) (2000), *Knowledge, language, and logic: Questions for Quine*. Boston Studies in the Philosophy of Science (pp. 407–30). Dordrecht: Kluwer Academic Publishers. Reprinted in Quine (2008b), pp. 259–68.

Quine, W. V. (2008a). *Confessions of a confirmed extensionalist and other essays*. Edited by D. Føllesdal & D. B. Quine. Cambridge, MA: Harvard University Press.

Quine, W. V. (2008b). *Quine in dialogue*. Edited by D. Føllesdal & D. B. Quine. Cambridge, MA: Harvard University Press.

Quine, W. V. & Ullian, J. S. (1970). *The web of belief*. Second edition (1978). New York: Random House.

Quinn, P. L. (1974). What Duhem really meant. In R. Cohen & M. Wartofsky (Eds.), *Methodological and historical essays in the natural and social sciences*. Boston Studies in the Philosophy of Science. Volume 14 (pp. 33–56). Dordrecht: Reidel.

Quinton, A. (1967). The importance of Quine. *New York Review of Books*, January 21.

Quinton, A. (1995). Analytic philosophy. In Honderich (1995), pp. 28–30.

Randall, J. H. (1944). Epilogue: The nature of naturalism. In Krikorian (1944), pp. 354–82.

Reck, A. J. (1968). *The new American philosophers: An exploration of thought since World War II*. Charlotte: Louisiana State University Press.

Reck, E. (Ed). (2013a). *The historical turn in analytic philosophy*. Houndmills: Palgrave Macmillan.

Reck, E. (2013b). Introduction: Analytic philosophy and philosophical history. In Reck (2013a), pp. 1–36.

Rescher, N. (1960). Review: *Word and object* by W. V. Quine. *American Scientist, 48*(4), 375A–377A.

Richardson, A. (1990). How not to Russell Carnap's *Aufbau*. In A. Fine, M. Forbers, & L. Wessels (Eds.), *PSA 1990: Proceedings of the Biennial Meeting of the Philosophy of Science Association*. Volume 1: *Contributed Papers* (pp. 3–14). East Lansing, MI: Philosophy of Science Association.

Richardson, A. (1992). Logical idealism and Carnap's construction of the world. *Synthese, 93*(1–2), 59–92.

Richardson, A. (1997). Toward a history of scientific philosophy. *Perspectives on Science, 5*(3), 418–51.

Richardson, A. (1998). *Carnap's construction of the world*. Cambridge: Cambridge University Press.

Ricketts, T. G. (1982). Rationality, translation, and epistemology naturalized. *Journal of Philosophy, 79*(3), 117–36.

Ricketts, T. G. (1985). Frege, the *Tractatus*, and the logocentric predicament. *Noûs, 19*(1), 3–15.

Romanos, G. (1983). *Quine and analytic philosophy*. Cambridge, MA: MIT Press.

Rorty, R. (Ed.). (1967a). *The linguistic turn: Essays in philosophical methods*. Chicago: University of Chicago Press.

Rorty, R. (1967b). Metaphilosophical difficulties of linguistic philosophy. In Rorty (1967a), pp. 1–40.

Rorty, R. (1970). Cartesian epistemology and changes in ontology. In Smith (1970), pp. 273–92.

Rorty, R. (1982). *Consequences of pragmatism*. Minneapolis: University of Minnesota Press.

Rorty, R. (1984). The historiography of philosophy: Four genres. In Rorty, Schneewind, & Skinner (1984), pp. 49–75.

Rorty, R., Schneewind, J., & Skinner, Q. (Eds.) (1984). *Essays on the historiography of philosophy*. Cambridge: Cambridge University Press.

Rosen, G. (1999). Review: *Naturalism in mathematics* by P. Maddy. *British Journal for the Philosophy of Science, 50*(3), 467–74.

Roth, P. (1999). The epistemology of "Epistemology naturalized." *Dialectica, 53*(2), 87–109.

Russell, B. (1900). *A critical exposition of the philosophy of Leibniz*. Second edition (1937). Cambridge: At the University Press.

Russell, B. (1914). *Our knowledge of the external world as a field for scientific method in philosophy*. Chicago: Open Court. Revised edition (1926). London: George Allen & Unwin.

Ryder, J. (1994). *American philosophic naturalism in the twentieth century*. Amherst, NY: Prometheus Books.

Ryle, G. (1949). *The concept of mind*. London: Hutchinson.

Salmon, W. & Wolters, G. (Eds.). (1994). *Logic, language, and the structure of scientific theories*. Proceedings of the Carnap-Reichenbach Centennial, University of Konstanz, May 21–24, 1991. Pittsburgh-Konstanz Series in the Philosophy and History of Science. Pittsburgh: University of Pittsburgh Press.

Santayana, G. (1923). *Scepticism and animal faith: Introduction to a system of philosophy*. New York: Charles Scribner's Sons.

Schilpp, P. (Ed.). (1963). *The philosophy of Rudolf Carnap*. Library of Living Philosophers. Volume 11. La Salle, IL: Open Court.

Schuldenfrei, R. (1972). Quine in perspective. *Journal of Philosophy, 69*(1), 5–16.

Sellars, R. W. (1922). *Evolutionary naturalism*. Chicago: Open Court.

Sellars, R. W. (1924). The emergence of naturalism. *International Journal of Ethics, 34*(4), 309–38.

Sellars, R. W. (1927). Why naturalism and not materialism? *Philosophical Review, 36*(3), 216–25.

Seth, A. (1896). The term "naturalism" in recent discussion. *Philosophical Review, 5*(6), 576–84.

Shapiro, S. (2000). The status of logic. In P. Boghossian & C. Peacocke (Eds.), *New essays on the a priori* (pp. 333–66). Oxford: Oxford University Press.

Shapiro, S. (Ed.). (2005). *The Oxford handbook of philosophy of mathematics and logic*. New York: Oxford University Press.

Shatz, D. (1994). Skepticism and naturalized epistemology. In Wagner & Warner (1993), pp. 117–45.

Siegel, H. (1995). Naturalized epistemology and "first philosophy." *Metaphilosophy, 26*(1), 46–62.

Sinclair, R. (2012). Quine and conceptual pragmatism. *Transactions of the Charles S. Peirce Society, 48*(3), 335–55.

Sinclair, R. (2014). Quine on evidence. In Harman and Lepore (2014), pp. 350–72.

Sinclair, R. (2016). On Quine's debt to pragmatism: C. I. Lewis and the pragmatic a priori. In Janssen-Lauret & Kemp (2016), pp. 76–99.
Sluga, H. (1980). *Gottlob Frege*. London: Routledge.
Smith, J. E. (1969). The reflexive turn, the linguistic turn, and the pragmatic outcome. *Monist, 53*(4), 588–605.
Smith, J. E. (Ed). (1970). *Contemporary American philosophy*. London: George Allen & Unwin.
Sober, E. (1999). Testability. *Proceedings and Addresses of the American Philosophical Association, 73*(2), 47–76.
Sober, E. (2000). Quine's two dogmas. *Proceedings of the Aristotelian Society Supplementary Volume, 74*(1), 237–80.
Spencer, H. (1862). *A system of synthetic philosophy: First principles*. London: Williams and Norgat.
Stenius, E. (1968). Beginning with ordinary things. *Synthese, 19*(1–2), 27–52. Reprinted in Davidson & Hintikka (1969), pp. 27–52.
Strawson, P. F. (1950). On referring. *Mind, 59*(235), 320–44.
Strawson, P. F. (1955). A logician's landscape. *Philosophy, 30*(114), 229–37.
Strawson, P. F. (1966). *The bounds of sense*. London: Methuen.
Strawson, P. F. (1983). *Scepticism and naturalism: Some varieties*. Woodbridge Lectures. London: Methuen (1985).
Stroll, A. (2000). *Twentieth-century analytic philosophy*. New York: Columbia University Press.
Stroud, B. (1968). Conventionalism and the indeterminacy of translation. *Synthese, 19*(1–2), 82–96. Reprinted in Davidson & Hintikka (1969), pp. 82–96.
Stroud, B. (1996). The charm of naturalism. *Proceedings and Addresses of the American Philosophical Association, 70*(2), 43–55. Reprinted in De Caro & Macarthur (2004a), pp. 22–35.
Tamminga, A. & Verhaegh, S. (2013). Katz's revisability paradox dissolved. *Australasian Journal of Philosophy, 91*(4), 771–84.
Tappenden, J. (2001). Recent work in the philosophy of mathematics. *Journal of Philosophy, 98*(9), 488–97.
Taylor, C. (1984). Philosophy and its history. In Rorty et al. (1984), pp. 17–30.
Tennant, N. (1994). Carnap and Quine. In Salmon & Wolters (1994), pp. 305–44.
Tennant, N. (2000). What is naturalism in mathematics, really? *Philosophia Mathematica, 8*(3), 316–38.
Thayer, H. S. (1967). Nagel, Ernest. In Edwards (1967), volume 6, pp. 440–41.
Thompson, M. (1964). Metaphysics. In R. M. Chisholm, H. Feigl, W. K. Frankena, J. Passmore, & M. Thompson (Eds). *Philosophy*, pp. 125–232. Englewood Cliffs, NJ: Prentice-Hall.
Urmson, J. O. (1961). The history of philosophical analysis. In Rorty (1967a), pp. 294–301.
Van Fraassen, B. (1996). Science, materialism, and false consciousness. In J. Kvanvig (Ed.), *Warrant in contemporary epistemology* (pp. 149–81). Lanham, MD: Rowman & Littlefield.
Verhaegh, S. (2014). Quine's argument from despair. *British Journal for the History of Philosophy, 22*(1), 150–73.
Verhaegh, S. (2015). Rafts, boats, and cruise ships: Naturalism and holism in Quine's philosophy. PhD dissertation. University of Groningen.
Verhaegh, S. (2017a). Boarding Neurath's boat: The early development of Quine's naturalism. *Journal of the History of Philosophy, 55*(2), 317–42.
Verhaegh, S. (2017b). Quine's "needlessly strong" holism. *Studies in History and Philosophy of Science, Part A, 61*(1), 11–20.
Verhaegh, S. (2017c). Quine on the nature of naturalism. *Southern Journal of Philosophy, 55*(1), 96–115.
Verhaegh, S. (2017d). Blurring boundaries: Carnap, Quine and the internal-external distinction. *Erkenntnis, 82*(4), 873–90.
Verhaegh, S. (2017e). Review: *Carnap, Quine, and Putnam on methods of inquiry* by G. Ebbs. *Notre Dame Philosophical Reviews*. November 23. Retrieved from http://ndpr.nd.edu/news/carnap-quine-and-putnam-on-methods-of-inquiry/.
Verhaegh, S. (2018a). Quine's dissatisfaction with "Two dogmas." Guest blog at Eric Schliesser's *Digressions and Impressions*. February 6. Retrieved from http://digressionsnimpressions.typepad.com/digressionsimpressions/2018/02/quines-dissatisfaction-with-two-dogmas-guest-post-by-sander-verhaegh.html.

Verhaegh, S. (2018b). Review: *Quine and his place in history* by F. Janssen-Lauret & G. Kemp (Eds.). *Philosophical Quarterly, 68*(271), 433–35.

Verhaegh, S. (forthcoming-a). *Sign and Object*: Quine's forgotten book project. *Synthese*. https://doi.org/10.1007/s11229-018-1693-z.

Verhaegh, S. (forthcoming-b). Setting sail: The development and reception of Quine's naturalism. *Philosophers' Imprint*.

Verhaegh, S. (forthcoming-c). "Mental states are like diseases": Behaviorism in the Immanuel Kant Lectures. In R. Sinclair (Ed.), *Science and sensibilia by W. V. Quine: The 1980 Immanuel Kant lectures*. London: Palgrave-Macmillan.

Verhaegh, S. (forthcoming-d). The behaviorisms of Skinner and Quine: Genesis, development, and mutual influence. *Journal of the History of Philosophy*.

Wagner, S. J. & Warner, R. (Eds.). (1993). *Naturalism: A critical appraisal*. Notre Dame: University of Notre Dame Press.

Wang, H. (1986). *Beyond analytic philosophy: Doing justice to what we know*. Cambridge, MA: MIT Press.

Ward, J. (1896). *Naturalism and agnosticism*. The Gifford Lectures delivered before the University of Aberdeen in the years 1896–1898. London: Adam and Charles Black (1899).

Watson, R. (1993). Shadow history in philosophy. *Journal of the History of Philosophy, 31*(1), 95–109.

Weir, A. (2005). Naturalism reconsidered. In Shapiro (2005), pp. 460–82.

Weir, A. (2014). Quine's naturalism. In Harman & Lepore (2014), pp. 114–47.

Wells, R. (1961). Review: *Word and object* by W. V. Quine. *Review of Metaphysics, 14*(4), 695–703.

White, M. (1948). On the Church-Frege solution of the paradox of analysis. *Philosophy and Phenomenological Research, 9*(2), 305–308. Reprinted (with postscript) in White (2005), pp. 112–15.

White, M. (1986). Normative ethics, normative epistemology, and Quine's holism. In Hahn & Schilpp (1986), pp. 649–62. Reprinted in White (2005), pp. 186–98.

White, M. (1999). *A philosopher's story*. University Park: Pennsylvania State University Press.

White, M. (2005). *From a philosophical point of view*. Princeton, NJ: Princeton University Press.

White, M. & Tarski, A. (1987). A philosophical letter of Alfred Tarski. *Journal of Philosophy, 84*(1), 28–32.

Wilder, H. T. (1973). Toward a naturalistic theory of meaning. PhD dissertation. University of Western Ontario.

Williamson, T. (2014). How did we get here from there? The transformation of analytic philosophy. *Belgrade Philosophical Annual, 27*(1), 7–37.

Yablo, S. (1998). Does ontology rest on a mistake? *Proceedings of the Aristotelian Society Supplementary Volume, 72*(1), 229–61.

Yeghiayan, E. (2009). The writings of Willard Van Orman Quine. *Rivista di Storia della Filosofia, 2009*(1), 187–238.

Yolton, J. W. (1967). *Metaphysical analysis*. London: George Allen & Unwin.

Yourgrau, W. (1961). Review: *Word and object* by W. V. Quine. *Zentralblatt fuer Mathematik, 53*(1), 3–4.

Zimmerman, D. (Ed.). (2008). *Oxford studies in metaphysics*. Volume 4. Oxford: Oxford University Press.

INDEX